FOOD SCIENCE AND TECHNOLOGY

# *CAJANUS CAJAN*

# CULTIVATION, USES AND NUTRITION

# Food Science and Technology

Additional books and e-books in this series can be found on Nova's website under the Series tab.

FOOD SCIENCE AND TECHNOLOGY

# *CAJANUS CAJAN*

## CULTIVATION, USES AND NUTRITION

DONALD S. WILKES
EDITOR

**Library of Congress Cataloging-in-Publication Data**

ISBN: 978-1-53619-134-9

*Published by Nova Science Publishers, Inc. † New York*

# CONTENTS

# PREFACE

Pigeon pea (*Cajanus cajan* (L.), among legumes, has an important role in the diet of many people in the world. It is one of the oldest food crops. It is the sixth most important legume crop. Pigeon pea is rich in protein, carbohydrates, and dietary fiber, and a rich source of other bioactive components. Pigeon pea is a good source of dietary fiber and is low in fat, which helps in the maintenance of body weight and reduces the risk of cardiovascular diseases. Cytoplasmic male-sterility (CMS) refers to the condition in plants where they fail to produce functional pollen.

In Chapter 1, the authors briefly discuss cytoplasmic-male sterility and its utilization in hybrid breeding in plants. Then they describe a historical overview of the discovery of male-sterility in pigeon pea. Next, a retrospective view on the major CMS systems developed and their use in commercial hybrid seed production in pigeon pea is presented. Finally, genomic approaches for stimulating pigeon pea hybrid breeding are briefly discussed.

In Chapter 2, the authors focus on the pharmacological and medicinal properties of pigeon pea. Next, the chemical composition of pigeon pea, its nutritional value, phytochemical components, health benefits and its usefulness in formulating functional foods is examined. In the final and fourth chapter, the cultivation, uses and other important nutritional information about this important legume is explored.

Chapter 1 - Cytoplasmic male-sterility (CMS) refers to the condition in plants where they fail to produce functional pollen. CMS occurs due to defects in the mitochondrial genome resulting in incompatibility with the nuclear genome. These defects result in unusual open reading frames (*orfs*) that may be chimeric and express altered proteins which obstructs the normal function of the mitochondria, thus disrupting pollen development and resulting in pollen sterility. Specific nuclear genes termed as a restorer of fertility (*Rf*) are competent in suppressing the male-sterile phenotype and restore the production of viable pollen in the plants with the aberrant mitochondrial genome. Cytoplasmic male-sterile systems have emerged as a reliable tool and extensively used in developing hybrids that enhances the yield of many crop plants. Pigeon pea is major grain legume crop grown primarily by the resource-poor farmers of the Indian sub-continent, South-East Asia and East Africa. Regarded for its moderately high protein content, minerals and vitamins and thus highly treasured from a food and nutritional security perspective. In the past three decades, spike growth has been recorded in the area and production of pigeon pea across the world, but its productivity has remained stagnant. To overcome this impediment in pigeon pea, extensive research has been performed and success was achieved by exploiting hybrid technology utilizing cytoplasmic male-sterile systems which can help in solving the seed production bottleneck. CMS has not been investigated in other legumes because of their self-pollinating nature, however pigeon pea is unique with considerable natural out-crossing. To accomplish the target of hybrid seed production, the key determinant is the availability of stable cytoplasmic male-sterile systems. So far, nine ($A_1$ to $A_9$) CMS systems have been developed in pigeon pea. Hybrids developed from cytoplasmic male-sterile systems have shown noteworthy yield improvement in pigeon pea under rain-fed conditions and minimal inputs.

Chapter 2 - Being rich in protein, dietary fibers and other nutrient contents, pulses are rightly called as poor people's meat and they make a well-balanced diet in combination with cereals. Pulses have several health benefits besides being nutritionally rich. These are contributed by large number and diversity of phytochemicals present as secondary metabolites like phenolics, phytates, lectins, flavonoids, saponins, phytosterols etc. which are reported to have pharmacological properties. Among legumes, *Cajanus cajan* (also commonly known as pigeon pea) is one of the most valuable, versatile, nutritional and health promoting grain crop of the world. Although primarily grown as an agricultural crop for food, it is the only woody legume in the world with affirmed medicinal properties including anti-inflammatory, anti-diabetic, anti-cancerous, antioxidant, hepato-protective, anti-helminthic etc. Its leaf extract can be used for treatment of skin infections, measles, jaundice and is also an antidote against constipation, food poisoning and stomatitis. In recent years, considerable progress and advances have been achieved in studying potential medicinal and biological applications of pigeon pea and establishing it as a medicinal drug. Different forms of aqueous, alcoholic as well as other organic solvent derived extracts from various tissues primarily leaves and root have been isolated to harness these phytochemicals. These have been known to act either directly or in combination with either other phytochemicals or trade drugs in alleviating certain disorders, diseases and are thus promising candidates for drug discovery. This chapter includes a detailed focused account of the presence of various organic phytocompounds in it such as stilbenes, flavonoids, polyphenols, cajaninstilbene acid, luteolin, apigenin, isorhamnetin etc. and the medicinal/pharmacological properties associated. With added advantage owing to its edibility in combination with excellent medicinal properties, *C. cajan* have the potential to be used as a pharmaceutical supplement in near future.

Chapter 3 - Pigeon pea (*Cajanus cajan* (L.), among legumes, has an important role in the diet of many people in the world. It is one of the oldest food crops. Is the sixth important legume crop, with more than 7 million hectares planted area and ~4.89 million tones productions. Pigeon pea is rich in protein, carbohydrates, and dietary fiber, and a rich source of other bioactive components. Pigeon pea is a good source of dietary fiber and is low in fat, which helps in the maintenance of body weight and reduces the risk of cardiovascular diseases. Moreover, there are various potential health benefits due to the phytochemicals, such as phenolics, flavonoids, phytates, lectins, tannins, saponins, oxalates, enzyme inhibitors and phytosterols. In addition, pigeon pea is also rich in minerals (iron, sulphur, calcium, potassium, and manganese), water-soluble vitamins (thiamine, riboflavin, niacin) and some indispensable amino acids, such as leucine, lysine, phenylalanine, isoleucine, and valine. On the other hand, pigeon pea has been used in the formulation and evaluation of various products such as creams to spread with *Lactobacillus reuteri,* pure with fruits and biscuits among others. It has also been used to replace wheat flour in various products. In this review, the authors discuss the chemical composition, nutritional value, phytochemical components, health benefits and its usefulness in formulating functional foods.

Chapter 4 - Pigeon pea [*Cajanus cajan* (L.) Millsp.] is a versatile pulse crop with high nutritive content. It can be grown in tropical and sub-tropical areas of the world. It can be cultivated in black cotton soils and well drained with a pH ranging from 7.0-8.5. As well as being a pulse crop pigeon pea fixes atmospheric nitrogen into soil and increases the fertility. India is major producer in the world and accounts for more than 70 percent of the production. Pigeon pea production is slowly increasing due to enormous breeding efforts like exploitation of heterosis by development of hybrids as well as male sterile lines and restorer lines in recent years. Pigeon pea is consumed as a raw immature seed, mature seed and processed dal. It is also used for animal feed and host material for lac production. Pigeon pea has also been exploited as a medicinal plant as the leaf extract contains an anti-microbial compound. It is an excellent source of protein (18-25%), starch (45-60%), crude fiber (8.2%) and fat (2.3%). It contains the best amino acid balance to fulfil daily nutrient requirement as well as minerals like calcium and magnesium in good amounts. Immature pigeon pea seed contains a high amount of vitamin A and vitamin C. Pigeon pea has excellent potential for the improvement of production and nutrient arability especially in poor developing countries due to its natural behavior.

In: *Cajanus cajan*
Editor: Donald S. Wilkes
ISBN: 978-1-53619-134-9


***Chapter 1***

# A Perspective on Cytoplasmic Male-Sterility Systems for Hybrid Breeding in Pigeon Pea

***Swati Saxena, Tanvi Kaila, Sandhya Sharma and Kishor Gaikwad****
ICAR- National Institute for Plant Biotechnology,
New Delhi, India

## Abstract

Cytoplasmic male-sterility (CMS) refers to the condition in plants where they fail to produce functional pollen. CMS occurs due to defects in the mitochondrial genome resulting in incompatibility with the nuclear genome. These defects result in unusual open reading frames (*orfs*) that may be chimeric and express altered proteins which obstructs the normal function of the mitochondria, thus disrupting pollen development and resulting in pollen sterility. Specific nuclear genes termed as a restorer of fertility (*Rf*) are competent in suppressing the male-sterile phenotype and restore the production of viable pollen in the plants with the aberrant

* Corresponding Author's E-mail: kish2012@nrcpb.org.

mitochondrial genome. Cytoplasmic male-sterile systems have emerged as a reliable tool and extensively used in developing hybrids that enhances the yield of many crop plants. Pigeon pea is major grain legume crop grown primarily by the resource-poor farmers of the Indian sub-continent, South-East Asia and East Africa. Regarded for its moderately high protein content, minerals and vitamins and thus highly treasured from a food and nutritional security perspective. In the past three decades, spike growth has been recorded in the area and production of pigeon pea across the world, but its productivity has remained stagnant. To overcome this impediment in pigeon pea, extensive research has been performed and success was achieved by exploiting hybrid technology utilizing cytoplasmic male-sterile systems which can help in solving the seed production bottleneck. CMS has not been investigated in other legumes because of their self-pollinating nature, however pigeon pea is unique with considerable natural out-crossing. To accomplish the target of hybrid seed production, the key determinant is the availability of stable cytoplasmic male-sterile systems. So far, nine ($A_1$ to $A_9$) CMS systems have been developed in pigeon pea. Hybrids developed from cytoplasmic male-sterile systems have shown noteworthy yield improvement in pigeon pea under rain-fed conditions and minimal inputs.

## 1. Introduction

Pulses reserve a significant position in Indian agriculture and within this legume group, pigeon pea (*Cajanus cajan* L.) is an economically and nutritionally important crop of the rain-fed and semi-arid regions of Asia, Africa, and Caribbean islands (Saxena et al., 2015). Pigeon pea belongs to the family *Fabaceae* and is a short-term perennial member, traditionally grown as an annual crop (Saxena et al., 2006). Primarily, it originated in India (Van der Maesen 1980), spreading to East Africa at least 3000 years ago, and then to Southeast Asia, West Africa, and the American continent. It is extremely popular for its high protein content (Saxena et al., 2002) and drought tolerance (Odeny 2007). Its seeds are a rich source of protein mainly fed on as dry dehulled splits or vegetables, the pods and foliage serves as a quality animal fodder while the dry woody stem contributes as fuelwood (Odeny 2007; Mallikarjuna et al., 2011). It is a preferred choice of resource-poor farmers largely in tropics and sub-tropics under low input environments (Saxena et al., 2005). Globally, pigeon pea is cultivated on over 7.03 million hectares with around 75% being produced in India and is highly treasured from a food and nutritional security perspective (FAO 2018). Pigeon pea is regarded as an ideal crop for substantial agriculture owing to its soil-enriching attributes such as fixation of atmospheric nitrogen, reuse of the soil nutrients, providing organic matter together with different essential minerals, and releasing soil-bound phosphorous (Saxena 2008).

In the past five decades, concerted efforts have been reported to enhance the production by developing promising high yielding varieties by employing hybridization programs. In spite of all these efforts, progress in the productivity has been limited and the improved cultivars fall short in enhancing the yield potential which has remained unchanged at around 700-800 kg $ha^{-1}$ (Saxena et al., 2018). This situation has raised serious concern taking into account the increasing population and declining per capita availability of proteins due to a stagnant production of the major pulse crops. Since pulses are the major source of protein in India and if the productivity of pigeon pea is not expanded significantly it will aggravate the issue of malnutrition specifically, among growing children (Saxena and Nadarajan 2010). Pigeon pea breeders have relentlessly worked towards breaking the yield barrier and this long-cherished goal was achieved by implicating a new hybrid breeding technology that has been profitably utilized in many crop species for accelerating the pigeon pea yield (Saxena et al., 2006).

Pigeon pea is distinctive among other legumes due to its floral morphology which allows partial natural out-crossing (Saxena et al., 1990). The hybrid pigeon pea breeding program was a joint initiative by the International Crops Research Institute for the Semi-arid Tropics (ICRISAT), Patancheru, Hyderabad, India and Indian Council of Agricultural Research (ICAR), New Delhi, India (Saxena and Nadarajan 2010). In 1991, the first-ever commercial genetic male-sterility (GMS) based pigeon pea hybrid, ICPH 8 was reported in India (Saxena et al., 1992). However, this hybrid was not commercialized owing to the high seed cost and tedious seed production method (Saxena et al., 2006). To overcome the above mentioned constraints, a more effective approach based on cytoplasmic male-sterile systems (CMS) was adopted for breeding hybrid pigeon pea. In pigeon pea, so far nine CMS ($A_1$ to $A_9$) systems have been bred leading to stable and fertile hybrids (Singh et al., 2017). The CMS-based pigeon pea hybrids have shown tremendous progress in increasing the productivity and break the long-standing yield barrier in this legume crop (Saxena et al., 2006).

In this chapter, we focus on the use of cytoplasmic male-sterility to produce hybrid seeds in pigeon pea. First, we briefly discuss cytoplasmic-male sterility and its utilization in hybrid breeding in plants. Then we describe a historical overview of the discovery of male-sterility in pigeon pea. Next, we present a retrospective view on the major CMS systems developed and their use in commercial hybrid seed production in pigeon pea. Finally, genomics approaches for stimulating pigeon pea hybrid breeding are briefly discussed.

## 2. Cytoplasmic Male-Sterility

In plants, cytoplasmic-male sterility (CMS) refers to impairment of the pollen formation which is an outcome of the complex interplay between the nuclear and the mitochondrial genomes. CMS leads to the formation of non-functional pollens, whereas the female gametophyte development remains unaltered (Schnable and Wise 1998). It is determined by the cytoplasmic factors and the phenotype is maternally inherited (Figure 1). CMS has been discovered in over 150 plant species (Laser and Lersten 1972). CMS is often associated with mutations in the mitochondrial genome which results in chimeric open reading frames (*orfs*) which in turn produce unique proteins causing mitochondrial dysfunction restricting the pollen development and ultimately concluding in male-sterility (Hanson and Bentolila 2004; Chase 2007). Specific nuclear genes known as *restorers of fertility* (*Rf*) "corrects" the defective pollen phenotype by suppressing cytoplasmic dysfunction and facilitates the formation of functional pollens subsequently restoring the male-fertility (Bentolila et al., 2002; Chase and Gabay-Laughnan 2004; Hanson and Bentolila 2004). The

CMS/*Rf* systems are valuable models for significantly improving our knowledge of cytoplasmic-nuclear communication (Eckardt 2006).

The phenotypic manifestations of male-sterility are quite different and is based on the nature of the plant species like (i) malformation or complete absence of stamens (male organ), (ii) deficit in the normal development of microspore tissues in anther, (iii) pollen abortion, (iv) inadequacy of the mature pollens to germinate on compatible stigma, (iv) formation of non-dehiscent anthers with functional pollens (Vinod 2005). CMS can emerge in the breeding populations in different ways (Hanson and Bentolila 2004) (i) spontaneously without any deliberate interference; (ii) inter-specific exchange between the cytoplasm and nuclear genomes; (iii) due to mutagenesis in the breeding lines; (iv) inter-specific hybridization between incompatible species (Table 2.1). In the past, CMS has been identified and described in many crop plants such as rice, wheat, onion, petunia, maize, onion, carrot, sunflower, sorghum, rye, pigeon pea, beet, *Brassica napus*, *Phaseolus vulgaris*, etc. (Table 2.2). Comprehensive in-depth analysis of CMS has been reported in excellent reviews by Schnable and Wise 1998; Budar and Pelletier2001; Chase 2007; Budar and Berthomé 2007; Chen and Liu 2014).

**Table 2.1. Different ways in which cytoplasmic male-sterility occurs**

| Occurrence | Crop species | CMS cytoplasm | Reference |
|---|---|---|---|
| Spontaneously | Maize (*Zea mays*) | CMS-Texas (CMS-T) | Roger and Edwardson 1952 |
| | *Brassica napus* | polima (pol) | Fu 1981; Liu et al., 1987 |
| | Rice (*Oryza sativa*) | CMS-WA (wild abortive | Seth et al., 1996 |
| Inter-specific exchange | Rice (*Oryza sativa*) | CMS- Boro II cytoplasm | Shinjyo 1966 |
| Mutagenesis | Sunflower (*Helianthus annuus*) | - | Horn et al.,1995 |
| Inter-specific hybridization | Sunflower (*Helianthus annuus*) | CMS-PET1 | Leclercq 1969 |

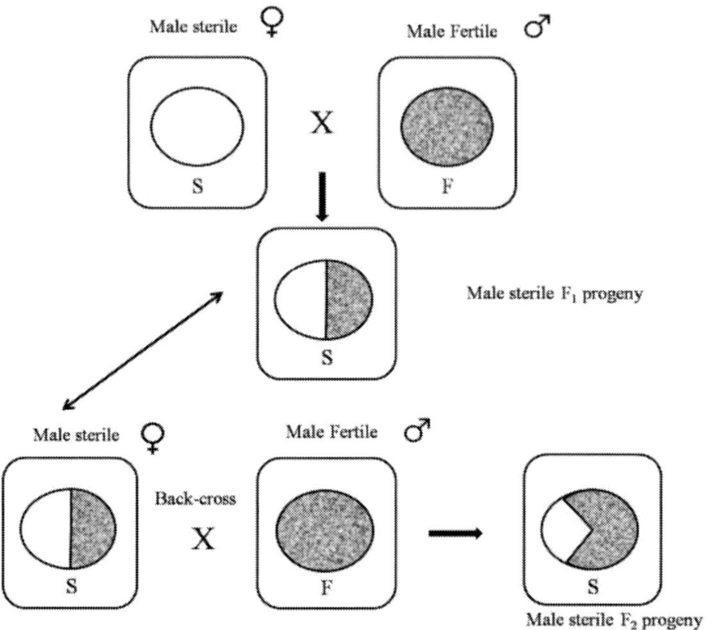


Figure 1. Maternal inheritance of cytoplasmic male-sterility. S denotes sterile cytoplasm and F denotes fertile cytoplasm. The progeny of male-sterile parents would always be male-sterile since cytoplasm of zygote comes primarily from the egg cell (Modified from Priyadarshan 2019).

## 2.1. History of Cytoplasmic-Male Sterility

Early evidence of cytoplasmic-male sterility was demonstrated in the 1900's (Duvick 1959; Havey 2004). Correns (1908) for the first time reported that in common summer savory (*Satureja hortensis* L.) all the hybrid progeny were male-sterile and maternally controlled. In 1921, Bateson and Gairdner, demonstrated that pollen-sterility in flax was maternally inherited from the female parent, while its expression was controlled by genes from either parent. Von Wettstein (1924) later attributed it to certain cytoplasmic (maternal) factors that were responsible for the phenotypic expression of male-sterility. Another report, on flax by Chittenden and Pellow (1927) reported that the male-sterility was due to the interplay between the cytoplasm and the nucleus. This phenomenon was an outcome of sterility-inducing cytoplasm and homozygous recessive nuclear genes. Rhoades (1931, 1931) stated that the male-sterility condition in corn (*Zea mays*) was inherited only via the female parent and the maternal cytoplasm was entirely responsible. All the progeny obtained were male-sterile with few occasional exceptions. Jones and Clarke (1943) published their report on onion and illustrated that the male-sterility trait was due to the interaction of male-sterile (S) cytoplasm and homozygous recessive (*msms*) nuclear genotype.

## 2.2. Cytoplasmic Male-sterility for Harnessing Hybrid Vigor

Hybrid vigor or heterosis is a phenomenon which results in the development of better performing $F_1$ progeny over their parent lines in terms of yield, better adaptability, improved stress tolerance (abiotic and biotic), and greater uniformity (Shull 1908; Saxena et al., 2013; Kim and Zhang 2018). During the 1860s, Charles Darwin originally described heterosis in maize (*Zea mays*). The use of heterosis has remarkably contributed in accelerating the yield gains of field crops, food security and has delivered immense economic profits in crop productivity globally (Whitford et al., 2013; Godfray et al., 2010; Tester and Langridge 2010). Majority of the important crop plants for instance maize, sorghum, rice, rapeseed, and sunflower have shown yield benefits from hybrid varieties (Tester and Langridge 2010). The pre-requisites for large-scale hybrid seed production are pollen transfer method and maintenance of the stable male-sterile plants which are used as maternal parents in hybrid crosses (Kim and Zhang 2018). Previously, methods like manual emasculation (removal of functional pollens) and chemical treatments were employed to develop male-sterile lines. However, these approaches were costly, tedious, and inefficient (Chen and Liu 2014). CMS is a widely adapted tool in hybrid breeding of various crop species as it circumvents the requirement for emasculation and ensures the production of male-fertile ($F_1$) progeny. The use of male-sterile lines expedites cost-effective commercial hybrid seed production (Mackenzie 2005).

**Table 2.2. List of cytoplasmic male-sterile systems and characterized CMS and restorer (*Rf*) genes in major crops (modified from Chen and Liu, 2014)**

| Crop species | CMS Cytoplasm | Associated CMS gene | Fertility restorer gene | References |
|---|---|---|---|---|
| Sugar beet (*Beta vulgaris*) | G | *nad9 and coxII* | *RfG1 and RfG2* | (Ducos et al., 2001) |
| | Owen | *preSatp6* | *Rf* | (Matsuhira et al., 2012; Yamamotoet al., 2005) |
| | I-12CMS | *orf129* | NA | (Yamamoto et al., 2008) |
| Brassica (*Brassica juncea*) | Hau | *atp6-orf288* | NA | (Wan et al., 2008) |
| | orf220 | orf220 | NA | (Yang et al., 2010; Zhang et al., 2003) |
| Brassica (*Brassica napus*) | Nap | *orf222/nad5c/orf139* | *Rfn* | (L'homme et al., 1997) |
| | Ogura | *orf138* | *Rfo* | (Bonhomme et al., 1992) |
| | Polima (pol) | *pol-orf (orf224)* | *Rfp* | (L'homme et al., 1997) |
| Brassica (*Brassica rapa*) | pol | *NA* | *Rfp1* | (Zhang et al., 2017) |
| Brassica (*Brassica tournefortii*) | Tour | *atp6-orf263* | *Rft* | (Landgren et al., 1996) |
| Wheat (*Triticum aestivum*) | AP | *orf256* | NA | (Song et al., 1994) |
| Sunflower (*Helianthus annuus*) | CMS3 | *coxIII and atp6* | NA | (Spassova et al., 1994) |
| | CMS89 | *orfH522, orfC* | NA | (Kohler et al., 1991, Laver et al., 1991) |
| | PET1 | *orfH522* | Rf1 | (Horn et al., 1996) |
| Perennial ryegrass (*Lolium perenne*) | MSL | *NA* | NA | NA |
| | Unknown | *atp6 and coxI* | NA | (Rouwendal et al., 1992) |
| | Unknown | *atp9* | NA | (McDermott et al., 2008) |
| Rye (*Secale cereale*) | Pampa | *pol-r* | *Rfp1 and Rfp2* | (Dohmen and Tudzynski 1994) |
| | Gulzow | *NA* | *Rfg1* | NA |

| Crop species | CMS Cytoplasm | Associated CMS gene | Fertility restorer gene | References |
|---|---|---|---|---|
| Rice (*Oryza sativa*) | Bo | *B-atp6* | *Rf1* | (Iwabuchi et al., 1993) |
| | BT | *atp6-orf79* | *Rf1 or Ifr1* | (Kazama et al., 2008) |
| | WA | *orf156, orfB* | *Rf4* | (Seth et al., 1996, Das et al., 2010) |
| | HL | *atp6-orfH79* | *Rf5 (Rf1a)* | (Hu et al., 2012; Wang et al., 2013) |
| | LD | *L-atp6-orf79* | *Rf2* | (Itabashi et al., 2011; Itabashi et al., 2009) |
| | CW | *orf307* | *Rf17* | (Fujii et al., 2007, 2010; Fujii and Toriyama 2009) |
| | RT120 | *rpl5-orf352* | *Rf102* | (Okazaki et al., 2013) |
| *Oryza sativa* | CMS-RT98 | *orf113-atp4-cox3* | NA | (Igarashi et al., 2013) |
| Petunia (*Petunia parodii*) | CMS3688 | *S-pcf* | Rf | (Hanson 1991) |
| Common bean (*Phaseolus vulgaris*) | Sprite | *pvs-atpA* | *Fr or Fr2* | (Johns et al., 1992) |
| Radish (*Raphanus sativa*) | Ogura | *atp6* | *Rfo* | (Makaroff et al., 1989) |
| | Kos | *orf125-atp8* | *Rfk1* | (Iwabuchi et al., 1999; Koizuka et al., 2003) |
| | Don | *orf463* | *Rfd1* | (Park et al., 2013) |
| Sorghum (*Sorghum bicolor*) | A3 (IS1112C) | *orf107* | *Rf3 and Rf4* | (Tang et al., 1999) |
| | Milo | *coxI* | *Rf1 and Rf2* | (Bailey-Serres et al., 1986) |
| | 9E | *coxI* | NA | (Bailey-Serres et al., 1986) |
| Maize (*Zea mays*) | C | *atp6, atp9 and coxII* | *Rf4, Rf5 and Rf6* | (Dewey et al., 1991) |
| | S | *orf355 or orf77* | *Rf3* | (Zabala et al., 1997) |
| | T | *T-urf13* | *Rf1 and Rf2* | (Dewey et al., 1987) |
| Pepper (*Capsicum annuum*) | Peterson | *cox2-orf456 and cox2-orf507* | NA | (Gulyas et al., 2010; Ji et al., 2013; Kim et al., 2007) |
| Carrot (*Daucus carota*) | Petaloid | *orfB* | NA | (Nakajima et al., 2001) |
| Barley (*Hordeum vulgare*) | msm1 | NA | *Rfm1 region* | (Ui et al., 2015) |

CMS has efficiently promoted the yield of many cereals, fruits, and vegetable crops by harnessing hybrid vigor or heterosis. During 1950's, the significance of hybrid vigor was practically exhibited by the development and utilization of hybrid corn. The maize CMS-T (Texas) system was the first cytoplasm system to be exploited for hybrid corn production leading to a remarkable six-fold increase in the maize yield between 1930 and 1990 in the U.S.A (Levings 1990). Following this success, CMS based-hybrid seed technology was executed in rice and other crops. In 1976, the first commercial hybrid rice was released in China, producing 8%-15% higher yield than the cultivated varieties. The hybrid rice varieties produced 12 tons per ha. Within a span of seven years (1998-2005), China released 34 "super" hybrid rice varieties producing an additional 6.7 million tons of rice (Priyadarshan 2019). The use of CMS for hybrid seed production has been extensively implicated in a plethora of major crop plants such as sorghum (*Sorghum bicolor* L. and *S. vulgare* L.), maize (*Zea mays* L.), rapeseed (*Brassica napus* L.), rye (*Secale cereale* L.), wheat (*Triticum aestivum* L.), pearl millet (*Pennisetum glaucum* L.), sunflower (*Helianthus annus* L.), castor (*Ricinus communis* L.), rice (*Oryza sativa* L.), and pigeon pea (*Cajanus cajan* L.) (Singh and Lohithaswa 2006; reviewed by Kaul 1988; Acquaah2012; Hariprasanna and Patil 2015; Zhao and Gai 2006; Geiger and Schnell 1970; Rajeshwari et al., 1994, Havey 2004; Dudhe et al., 2011; Ingale et al., 2006; Barclay 2010; Das et al., 2014; Saxena et al., 2011, 2013).

The three-line hybrid breeding system for hybrid seed production has been well established in many crop plants (Chen and Liu 2014). It involves the CMS (A) line, the maintainer (B) line, and the restorer (R) line (Figure 2). The CMS (A) line, contains male-sterile cytoplasm and recessive nuclear restorer (*rf*) genes. The maintainer (B) line, carries a similar nuclear genome (*rf*) as that of the CMS line but has normal fertile cytoplasm that allows plants to be fertile. The B-line is utilized to maintain the sterile line and termed as the maintainer line. The restorer (R) line possesses fertile cytoplasm along with dominant *Rf* genes. The B-line and the R-line produce the seeds by self-pollination. The A-line is used as a female parent and the R-line is used as the male parent in the crosses to produce $F_1$ hybrids. The $F_1$ hybrids developed possess the dominant *Rf* gene to restore male-fertility and exhibit substantially improved performance (i.e., hybrid vigor).

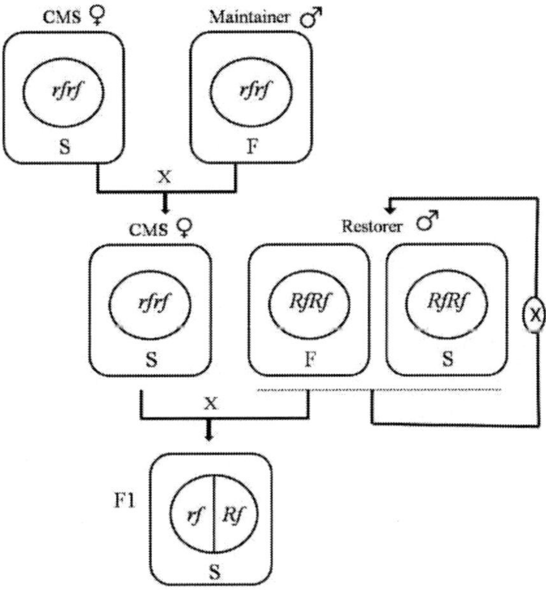


Figure 2. Three-line hybrid breeding system. It involves a CMS line, with sterile cytoplasm (S) and a non-functional recessive restorer (*rf*) gene(s); a maintainer line, with normal cytoplasm (F) and non-functional recessive restorer (*rf*) gene; and a restorer line, with normal (F) or sterile (S) cytoplasm and a functional dominant restorer (*Rf*) gene(s). (Modified from Chen and Liu 2014).

## 2.3. Genes Involved in Cytoplasmic Male-Sterility and Fertility Restoration

Cytoplasmic male-sterile systems have proved to be perfect for commercial hybrid seeds production of field crops. Genetic factors carried by both the parents (male and female) partially control the CMS expression and the same is retained over generations. The mitochondrion is a semi-autonomous body as it possesses its own DNA and can replicate by itself. It also synthesizes its own essential proteins related to the TCA cycle and

ATP generation. Mostly, mitochondrial genes, governing cytoplasmic male-sterility have been determined as *orfs* (open reading frames) obtained from the fusion of mitochondrial gene-coding and flanking sequences and sequences of unidentified origin (Chase and Laughnan 2004; Hanson and Bentolila 2004). With the help of recombination events occurring in plant mitochondria, these sequences get fused with the promoter sequence or are placed downstream of a sequence that supports gene expression and is thus expressed (Chase 2007). Therefore, it can also be said that CMS causing genes appears as parasitic sequences which uses expression signals of essential mitochondrial genes necessary for cell survival (Budar and Pelletier 2001). In most of the cases, such *orfs* which determine CMS have been observed to be gain of function mutation except, few cases where the loss of function mutation has resulted in CMS (Ducos et al., 2001; Sabar et al., 2000). In most of the cases, CMS causing *orfs* were found to be associated with the electron transfer chain of mitochondria and genes encoding transmembrane proteins. The mechanism of chimeric *orfs* resulting in CMS has also been reported for some species. In *Raphanus sativa* L., *orf138* was found to be associated with inhibition of CHS (chalcone synthase) before bud formation, thus, resulting in CMS phenotype in Ogura CMS of *Raphanus sativa* L. (Yang et al., 2008). In another example, modified gene expression of the *atp6* subunit of mitochondrial ATP synthase was responsible for CMS in sorghum (Howard and Kempken 1997). There are various strategies that can be applied to detect candidate CMS genes but the commonly used approach is the comparison of transcriptome or proteome of CMS line and fertility restored line of a crop plant. Liu et al., 2007 used this approach and identified *WA352* as the candidate gene causing CMS in the rice CMS-WA line. Similarly, by comparing proteome of sterile and fertile lines, URF13 of maize CMS-T (Forde et al., 1978) and truncated COX2 of sugar beet CMS-G (Ducos et al., 2001) were reported as candidate proteins causing CMS. Transferring a recombinant construct carrying CMS gene and 5’ mitochondrial targeting signal sequence to the nuclei of various plants (rice, tobacco, rapeseed, mustard and sugar beet) has helped in the verification of candidate CMS genes (Chen and Liu 2014).

The *Rf* genes (restorers of fertility) are nuclear-encoded genes that suppress CMS causing genes (Chase 2007). Most of the *Rf* genes belongs to the PPR (Pentatricopeptide Repeat) protein family. The PPR proteins are represented by a tandem array of degenerate amino acid repeat sequences (Okuda et al., 2007). Extensive studies have concluded that PPR proteins play an important role in post-transcriptional regulation *viz.* RNA editing by binding to the target transcript and hence, stabilizing the RNA. This might result in the formation of a new start and stop codons, thereby, changing the length of CMS causing *orfs* (Schnable and Wise 1998). It has been reported in sorghum that, editing of the *atp6* gene of mitochondria enhances fertility restoration (Howard and Kempken1997). However, as an exception to the above-mentioned fact, the *Rf2* gene encoding aldehyde dehydrogenase and *Rf1* gene encoding putative peptidase of the M48 family have been reported to restore fertility in maize CMS-T and sugar beet CMS-Owen (Hagihara et al., 2005; Matsuhira et al., 2012), respectively indicating alternate pathways of fertility restoration. As far as fertility restoration is concerned, in some cases, a single *Rf* gene is reported to be responsible for fertility restoration (WA cytoplasm of rice, Pol cytoplasm of rapeseed, Ogura cytoplasm of radish) (Ahmadikhah and Karlov 2006; Singh and Brown 1993; Koizuka et al., 2003). While in others, two or more than two *Rf* genes may be involved in fertility restoration (CMS-T and CMS-C cytoplasm of maize, PET-1 cytoplasm of sunflower and T-cytoplasm of onion) (Levings and Dewey 1988; Sotchenko et al., 2007; Levings 1993, Schnable and Wise 1998). A list of candidate CMS genes and fertility restorer genes reported in major crops is provided in Table 2.2.

## 3. Early Research in Breeding Pigeon Pea Hybrids

Stagnant production, high prices, and spike increase in imports of pigeon pea have raised serious concern to the prime stakeholders in India (Saxena and Nadarajan 2010). To break the yield barrier and increasing pigeon pea production has been a long-cherished goal of breeders. A breakthrough, by developing hybrid breeding technology has provided the needed respite. In the past, a tremendous leap in the yield of some cereals, fruits, and vegetables has been observed by commercially exploring heterosis or hybrid vigor (Saxena et al., 2006). The event of heterosis has been widely explored in cross-pollinating crop plants and suitable breeding methods were developed to produce improved hybrid varieties (Saxena 2002). In legumes, out-crossing is negligible, whereas pigeon pea is an exception with partial insect-aided natural out-crossing and its extent varies from one environment to another (Saxena et al., 1983). In pigeon pea, natural out-crossing was first reported by Howard et al., (1919). However, in the past, the most breeder ignored this fact and pigeon pea was manipulated as a self-pollinated crop in the breeding programs. The discovery of male-sterility marked the advent of hybrid pigeon pea technology (Saxena and Nadarajan 2010).

### 3.1. Genetic-Male Sterility (GMS) in Pigeon Pea

Kolreuter (1763) for the first time reported male-sterility in plants. Genetic-male sterility (GMS) in plants refers to another type of reproductive abnormality where they fail to produce functional pollens and this condition is governed by a single recessive nuclear (*ms*) gene (Vedel et al., 1994). In nature, GMS happens as a result of alteration of the male-fertility dominant gene to its recessive state due to some natural forces. In, self-pollinating crop plants, these genotypes are inevitably lost, compared to the out-crossing (fully or partial) crops, where they appear within the population in heterozygote form (Saxena et al., 2010). Based on the reports so far, all the GMS systems identified in pigeon pea have emerged due to spontaneous mutations.

Deshmukh (1959) gave the first report on male-sterility in pigeon pea, however, due to its tight linkage with female-fertility they could not be maintained. In 1974, ICRISAT started the hybrid breeding program to utilize the natural out-crossing in pigeon pea for enhancing crop production. Systematic efforts were made to search male-sterility systems in the germplasm and success was achieved when Reddy et al., (1978) identified translucent anthers in ICP 1596 accession. The male-sterility was controlled by a single recessive gene (*$ms_1$*) and it was reported as a stable GMS system. In 1983, Saxena and coworkers identified a second stable source of GMS system governed by a single (*$ms_2$*) recessive gene and represented by brown arrow-head shaped anthers. The prominent anther morphology of these two GMS systems facilitates easy identification of the male-sterile plants in the fields (Saxena 2002). Some other GMS system were also reported in pigeon pea and are listed below in Table 3.1 (see Saxena et al., 2010).

**Table 3.1. Genetic male-sterile (GMS) systems in pigeon pea (modified from Saxena et al., 2010)**

| Reported by | Characteristics | Genes |
|---|---|---|
| Deshmukh (1959) | Male-sterility associated with female sterility | – |
| Reddy et al., (1978) | Translucent male-sterile anthers | $ms_1$ |
| Venkateshwara et al., (1981) | Male-sterility related with obcordate leaf type | – |
| Saxena et al., (1981) | Partial male-sterility with limited pollen formation | – |
| Dundas et al., (1982) | Photo-insensitive male-sterile system | – |
| Saxena et al., (1983) | Brown arrow-head shape anthers | $ms_2$ |
| Gupta and Faris (1983) | Small white non-dehiscent anthers, which later turns brown; recessive gene control | – |
| Pandey et al., (1994) | Male-sterility related with obcordate leaf type | – |
| Verulkar and Singh (1997) | Translucent male-sterile anthers, limited podding and delay in flower induction; single recessive gene control | – |
| Wanjari et al., (2000) | Single dominant gene control | – |
| Saxena and Kumar (2001) | Small light yellow color anthers without pollen grains | $ms_3$ |

## 3.2. GMS-Based Pigeon Pea Hybrids

The availability of GMS sources in pigeon pea laid the foundation for developing commercial hybrid seed technology. The GMS system controlled by the $ms_1$ recessive gene (Reddy et al., 1978) was extensively utilized for developing hybrids to evaluate the degree of hybrid vigor and their ability to out-cross (Saxena et al., 2015). Initially, the GMS source by

Reddy et al., (1978) was used for developing 53 experimental hybrids, of which ten hybrids displayed 20-40% hybrid vigor. In 1990, more enhanced GMS sources were developed and within a time frame of 8 years, 203 hybrids were evaluated and all the hybrids displayed more than 20% standard heterosis. Of these 203 hybrids, 80 and 46 hybrids exhibited >40% and >80% heterosis over the control plants, respectively. This revealed valuable knowledge about exploitable heterosis in pigeon pea (Saxena et al., 2018). Later, ICRISAT collaborated with ICAR and they invested enormous resources to achieve success in this endeavor. In 1991, the joint effort resulted in a massive breakthrough with the release of the world's first pigeon pea hybrid ICPH8. It was developed by crossing the GMS line (MS Prabhat DT) with a fertile inbred line ICPL 161. Since no commercial hybrids were available in legumes, ICPH 8 is regarded as a breakthrough in the history of pulse breeding. The hybrid ICPH8 was evaluated in 100 field trials where it performed exceptionally well and out-performed the controls UPAS120 and Mayank by 41% and 34.2%, respectively (Saxena et al., 1992). Later, five more GMS-based hybrids were bred and released by ICAR, India. In 1993 a short-duration hybrid PPH 4 was released by Punjab Agricultural University (PAU) Ludhiana demonstrating superiority by 14% (Verma and Sandhu 1995). In 1994 another short-duration hybrid CoH 1 was released by Tamil Nadu Agricultural University (TNAU), Coimbatore which displayed a 32% increased yield than the control variety (Murugarajendran et al., 1995). Three years later, in 1997, TNAU released a hybrid variety CoH 2 which exhibited higher yield over the CoH 1 and the control by 13% and 35%, respectively. Later, in 1997 and 1998, two more pigeon pea hybrids AKPH4104 and AKPH2022 were released by Dr. Punjabrao Deshmukh Krishi Vidyapeeth (PDKV), Akola. The hybrids out-yielded their controls by 64% and 35%, respectively (Wanjari et al., 1999). Details of the GMS-based hybrids are provided in Table 3.2.

**Table 3.2. Summary of GMS-based pigeon pea hybrids developed in India (modified from Saxena 2002 and Saxena et al., 2006)**

| Hybrid | Year of Release | Adaptability | Parents | Days to maturity | Yield (kgha$^{-1}$) | Superiority over control |
|---|---|---|---|---|---|---|
| ICPH 8 | 1991 | Central zone | MS Prabhat DT x ICPL 161 | 125 | 1780 | 41% over UPAS |
| PPH 4 | 1993 | Punjab | MS Prabhat DT x AL 688 | 137 | 1930 | 14% over UPAS, 14% over H 82-1 |
| CoH 1 | 1994 | Tamil Nadu | MS T 21 x ICPL 87109 | 117 | 1210 | 22.3% over Vamban-1 |
| CoH 2 | 1997 | Tamil Nadu | MS Co 5 x ICPL 83027 | 120-130 | 1050 | 35% over Co 5 |
| AKPH 4104 | 1997 | Central zone | NA | 130-140 | NA | 64% over UPAS |
| AKPH 2022 | 1998 | Maharashtra | NA | 180-200 | NA | 35% over BDN 2, 25% over ICPL 87119 |

NA- Data not available.

### 3.3. Adoption of GMS-Based Pigeon Pea Hybrids

The released GMS-based hybrids were high-yielding and displayed superiority over the control varieties in multi-location assessment performed across the country. In spite of these attributes, the GMS-based hybrids were not accepted by the farmers, public, and private seed companies. The main issue was the maintenance of the purity of the female parent as well the hybrid seed as its execution resulted in increased production value. The manual removal of the fertile plants (50%) in reach of the female rows caused practical constraints in seed production because it was tedious, time-consuming, and labour-intensive (Saxena et al., 2006). Niranjan et al., (1998) assessed the reason behind the downfall and established the problems encountered by the farmers, researchers, and seed companies in the adoption of hybrid seed technology. They stated that in spite of the low and economical production cost of the GMS-hybrid seed and the profitable hybrid dominance, still the technology itself suffered from serious impediment when it comes to large-scale seed production. (Saxena 2002; Saxena et al., 2006; Saxena and Nadarajan 2010). The overall experience with the GMS-based hybrids, provide two valuable facts (i) availability of sufficient heterosis in pigeon pea; (ii) setting of adequate pods due to available natural out-crossing (25%) (Saxena and Nadarajan 2010).

## 4. Cytoplasmic Male-Sterile Systems in Pigeon Pea - A Game Changer

In pigeon pea, the limitations encountered in large-scale seed production inherited with GMS-based hybrids were efficiently tackled by employing more systematic hybrid technology based on cytoplasmic male-sterile (CMS) systems. The CMS systems occurs due to spontaneous mutation, intra-specific cross, inter-specific cross or inter-generic cross (Saxena et al., 2010). To date, the wide hybridization method involving inter-specific, or inter-generic cross has revealed successful results in developing CMS systems in various crop plants (Kaul 1988). The CMS systems provide abundant opportunities to utilize hybrid vigor in pigeon pea.

## 4.1. Initial Efforts to Develop CMS Lines

Reddy and Faris (1981) performed the preliminary attempt to develop a CMS line in pigeon pea by crossing a cultivated type (as a female parent) with pollens from two wild relatives, *C. scarabaeoides* and *C. sericeus*. The $F_1$ progeny obtained from both the crosses were fertile and were further utilized as a male partner to generate backcrosses with wild relatives as female partners. The $BC_1F_1$ progeny were male-fertile while some $BC_1F_2$ plants displayed maternally inherited male-sterility. However, this male-sterility was linked to various floral deformities like free stamen, petaloid anthers, or heterostyly. Therefore, these CMS lines were unstable and not employed in hybrid breeding (Saxena et al., 2018).

Ariyanayagam et al., (1993) made the following attempt by employing chemical and physical mutagens. They treated a GMS line (controlled by $ms_2$) with 0.025% sodium azide or 500 mg/kg streptomycin sulphate which displayed some mutations and expressed maternally inherited male-sterility. The partial male-sterile lines obtained in this effort were not stable and failed to establish a functional CMS system.

## 4.2. Development of CMS Systems - The Real Breakthrough

After the unsuccessful attempts, the real breakthrough was accomplished by integrating the nuclear genome of the cultivated type in the wild-relative cytoplasm through interspecific-hybridization followed by several rounds of back-crossing (Figure 3). It was believed that this strategy would give the desired results by producing male-sterile lines (Saxena 2013). Keeping this in view, concerted efforts were made in pigeon pea and nine CMS-inducing cytoplasmic ($A_1$ to $A_9$) systems have been bred (Table 4.1). These CMS systems contribute to developing a more robust and effective hybrid-breeding program. The CMS sources developed in pigeon pea are briefly discussed below.

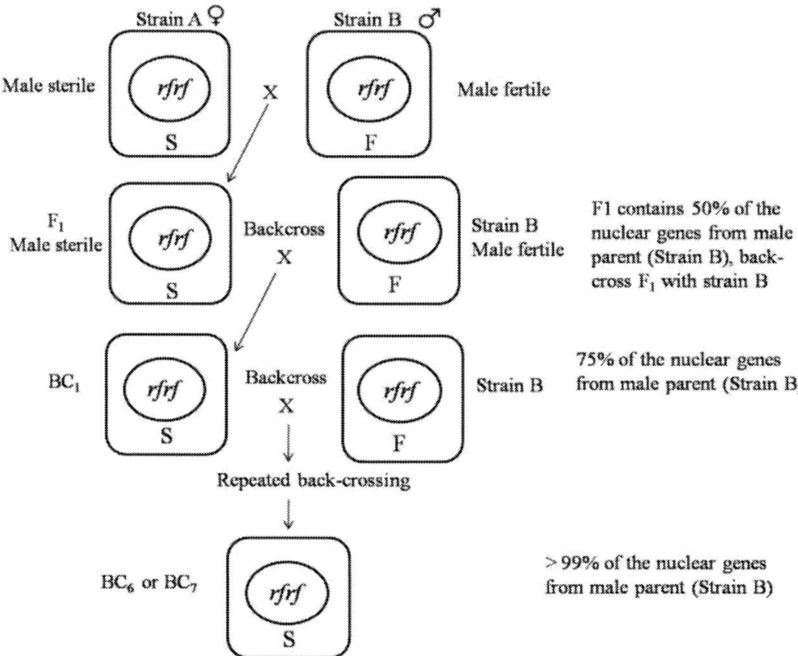


Figure 3. Breeding procedure for transferring of male-sterile cytoplasm from strain A to strain B. Strain A (male-sterile) is used as the non-recurrent parent in a back-cross approach with strain B (male-fertile). After repeated 6-7 back-crosses the progeny will be similar to strain B in the nuclear genotype. The strain B serve as the maintainer line of the new male-sterile strain. (Modified from Singh 2005).

### *4.2.1. $A_1$ CMS System with Cajanus sericeus (Benth. ex Bak.) van der Maesen Cytoplasm*

Ariyanayagam et al., (1993) performed the first attempt by crossing *Cajanus sericeus* (as a female parent) with a cultivated genotype (as a male parent). The $F_1$ progeny displayed partial male-sterility, whereas, in the $F_2$ progeny, few segregates were identified with 100% pollen sterility. The male-sterile lines cannot be maintained as some of them showed reversion to male-fertility due to local environmental fluctuations. Ariyanayagam et al., (1995) performed an alternative approach of multiple genome transfer to stabilize the male-sterility. Few male-sterile plants were derived by this approach following their maintenance by using pigeon pea inbred lines. These male-sterile lines were processed further through additional hybridization and new CMS lines (CMS 85010A, CMS 88034A, and CMS 13091A) were developed (Saxena et al., 1996). This cytoplasm was identified as $A_1$. This $A_1$ cytoplasm showed temperature sensitivity. The CMS lines reverted to male-fertility under low temperatures and again regained male-sterility under high temperatures. This trait makes $A_1$ cytoplasm unsuitable for exploiting heterosis (Mallikarjuna et al., 2012).

**Table 4.1. The summary of CMS systems obtained from different wild relatives of pigeon pea**

| CMS system | Wild relative (cytoplasm donor) | Gene pool | References | Remarks |
|---|---|---|---|---|
| $A_1$ | *C. sericeus* | Secondary | Ariyanayagam et al., (1993); Saxena et al., (1996) | Sensitive to temperature, unstable |
| $A_2$ | *C. scarabaeoides* | Secondary | Tikka et al., (1997); Saxena and Kumar (2003) | Used in hybrid breeding |
| $A_3$ | *C. volubilis* | Tertiary | Wanjari et al., (1999) | No restorers found |
| $A_4$ | *C. cajanifolius* | Secondary | Saxena et al., (2005a) | Stable and employed in hybrid breeding |
| $A_5$ | *C. cajan* | Primary | Mallikarjuna and Saxena (2002) | – |
| $A_6$ | *C. lineatus* | Secondary | K.B. Saxena (unpublished data) | – |
| $A_7$ | *C. platycarpus* | Tertiary | Mallikarjuna et al., (2006) | – |
| $A_8$ | *C. acutifolius* | Secondary | Saxena (2013) | – |
| $A_9$ | *C. cajan* | Primary | Srikanth et al., (2015) | – |

Source: Saxena et al., 2018.

#### 4.2.2. $A_2$ CMS System with *Cajanus scarabaeoides* (L.) Thou. Cytoplasm

Ariyanayagam et al., (1993) performed the preliminary attempt by using *C. scarabaeoides* (as female) and crossed it with pigeon pea (ICPL 85030) line. The resulting $F_1$ progeny showed partial male-sterility and the backcrosses showed few favorable male-sterile segregates, however stable CMS lines could not be obtained. Tikka et al., (1997) developed the CMS line by integrating the cytoplasm of *C. scarabaeoides* with the genome of a cultivated type. They obtained male-sterile plants in the $F_2$ generation, along with an ideal maintainer line (ICPL 288). Later, fertility restorers of the CMS line were detected among the $F_2$ generation and the cytoplasm was named as $A_2$. The $A_2$ CMS system displayed increased stability across different environments. It was practically implemented for producing hybrids in Gujarat, India.

In a subsequent study, *C. scarabaeoides* (as female) was crossed with four pigeon pea cultivars (Saxena and Kumar 2003). The cross between *C. scarabaeoides* x ICPL 88039 produced 100% male-sterile $F_1$ plants. The male-sterile system was stabilized by repeated backcrossing with ICPL 88039 (as the recurrent parent), ultimately producing complete male-sterile lines. They reported many sources of maintainers and restorers. This CMS system showed variations in pollen-fertility. Later, fertility restoration was found among $F_2$ segregants and a good maintainer line was also identified (Saxena et al., 2010).

#### 4.2.3. $A_3$ CMS System with *Cajanus volubilis* (Blanco) Blanco Cytoplasm

Wanjari et al., (1999) performed a cross between *C. volubilis* (as the female parent) with a cultivated type and reported the identification of male-sterile segregates. This cytoplasm was designated as $A_3$. The lack of availability of stable fertility restorers, discouraged the use of this cytoplasm for a hybrid breeding program.

#### 4.2.4. $A_4$ CMS System with *Cajanus cajanifolius* (Haines) Maesen Cytoplasm

Rathnaswamy et al., (1999) attempted to cross *C. cajanifolius* (male parent) with a GMS line (female parent) but failed to breed a CMS source. While, *C. cajanifolius*is placed in the secondary gene pool, it is regarded as the immediate wild relative to the cultivated group (Ratnaparkhe et al., 1995). They both differ only by 1-5 genes (Mallikarjuna et al., 2012) and *C. cajanifolius* is proposed as the progenitor of the cultivated pigeon pea (De 1974; van der Maesen 1980). A cross was performed between *C. cajanifolius* (as female) and cultivate pigeon pea ICPL 28 (Saxena et al., 2005). The $F_1$ hybrids were fully male-sterile with absence of pollens, exhibited perfect female gametophyte, and showed no signs of morphological deformity (Saxena et al., 2010). This cytoplasm was named as $A_4$ and is the best amongst the other identified CMS sources. This CMS system along with its fertility restorers is highly stable across various environments (Chaudhari et al., 2015; Sawargaonkar et al., 2012a) and is extensively utilized for breeding commercial pigeon pea hybrids.

#### 4.2.5. $A_5$ CMS System with *Cajanus cajan* (L.) Millsp. Cytoplasm

The first attempt was made by Rathnaswamy et al., (1999), by crossing a GMS line (female parent) with *Cajanus acutifolius* (F. V. Muell.) van der Maesen (male parent) resulting in complete male-fertile $F_1$ progeny. Later, *C. acutifolius* (female parent) was crossed with pigeon pea accession ICP 1140 (male parent) (Mallikarjuna and Saxena 2002). Gibberellic acid (at 50 mg/L) was applied to increase pod set, however the hybrid seed obtained was underdeveloped and did not germinate. Mallikarjuna and Moss (1995) tried to defeat this constraint, by recovering the developing embryos and effectively cultured them in artificial medium. However, no success was achieved in developing a CMS source.

Mallikarjuna and Saxena (2005) used the embryo rescue method, to cross six pigeon pea cultivars (as female) with two accessions of *C. acutifolius* (ICPW 15613, ICPW 15605). The $F_1$ generations of three pigeon pea cultivars (ICPL 85010, ICPL 85030, and ICPL 88014) showed male-sterile segregates, with few of them displaying 100% pollen sterility.

The anthers of the $F_1$ plants were shriveled and pale yellow in appearance. The pollen-sterility was maintained by crossing with their corresponding wild relative. The majority of the cultivated types were capable in restoring the male-fertility except HPL 24. The $F_1$ generation contained both male-sterile and male-fertile plants, indicating the existence of both *rf* and *Rf* nuclear genes (Saxena et al., 2013). Many fertile maintainers were also identified and backcrossing the male-sterile lines with their fertile maintainer resulted in the stable CMS system (K. B. Saxena unpublished data). The availability of good stable maintainers and restores, implies the use of $A_5$ cytoplasm for exploiting hybrid vigor in pigeon pea (Saxena et al., 2010).

#### *4.2.6. $A_6$ CMS system with Cajanus lineatus (W & A) van der Maesen cytoplasm*

During the year 2002, a naturally cross-pollinating partial male-sterile plant with a unique morphology was detected in a population of *C. lineatus* (K. B. Saxena unpublished data). The vegetative cuttings of the same was grown in a glasshouse and upon examination, it was observed that all the plants were male-sterile. The cross between the male-sterile line and pigeon pea accession ICPL 99044 resulted in partially male-sterile $F_1$ progeny. In a back-cross ($BC_1F_1$) with ICPL 99044, five out of the twenty plants raised showed partial male-sterility. In $BC_4F_1$ generation, 167 plants were studied for male-sterility and the results varied between 92%-100%. The four plants with complete male-sterility were crossed with ICPL 99044 for sustaining the cytoplasmic male-sterility (Saxena et al., 2018).

#### *4.2.7. $A_7$ CMS with Cajanus platycarpus (Benth.) van der Maesen Cytoplasm*

A wild type *C. platycarpus*, belongs to the tertiary gene pool and is cross-incompatible with the cultivated pigeon pea. Mallikarjuna et al., (2006), therefore, performed hormone-assisted pollinations associated with embryo rescue technique to produce functional $F_1$ and $BC_1F_1$ generation. Within the $BC_2F_1$ generation, a $BC_2$-E progeny exhibited low pollen-fertility. In this progeny, two segregates with complete pollen-sterility were picked and crossed with few pigeon pea cultivars (ICPL 85010, ICPL 88014 and ICP 14444). The $F_1$ hybrids involving ICPL 85010 cultivar were completely male-sterile. On the other hand, the cultivars ICPL 88014 and ICP 14444 were successful in restoring male-fertility in the hybrids (Mallikarjuna et al., 2011).

#### *4.2.8. $A_8$ CMS System with Cajanus reticulatus (Aiton) F. Muell Cytoplasm*

At ICRISAT, an out-crossed plant having unique morphological features was detected amid the population of *C. reticulatus*. In this plant, pollen development was limited and it displayed partial male-sterility. The cross between this plant and an inbred cultivar resulted in a fully male-sterile $BC_2F_1$ generation. The plants of the $BC_2F_1$ progeny had underdeveloped androecium which failed to produce functional pollen. Till now six good maintainers are reported and exploration for stable fertility restorers is underway (Saxena 2013).

#### *4.2.9. $A_9$ CMS System from Cajanus cajan (L.) Millsp. Cytoplasm*

Recently, the $A_9$ CMS system has been developed and is based on cultivated cytoplasm. In the $A_5$ cytoplasm, *C. acutifolius* was used as a male parent, however in the $A_9$ cytoplasm, *C. lanceolatus* (W. Fitzg) Maesen was the preferred male counterpart (Shrikanth et al., 2015). The availability of maintainers and restorers of the $A_9$ CMS system makes it a preferable choice in comparison to the $A_5$ CMS system (Saxena et al., 2018).

Although, the CMS ($A_1$-$A_9$) systems developed in pigeon pea exhibit a broad-scope of mitochondrial variability, till now only two ($A_2$ and $A_4$) systems have been utilized for commercial hybrid breeding due to the availability of perfect maintainers and fertility restorers and their stability under various environmental conditions (Saxena and Nadarajan 2010). This situation makes it imperative to perform focused research on breeding new diverse CMS sources/systems exhibiting both cytoplasmic and nuclear diversity (Saxena et al., 2018).

# 5. Cytoplasmic Male-Sterility Based Pigeon Pea Hybrids

The discovery of cytoplasmic male-sterility was a major milestone in the development of hybrid pigeon pea. Soon after the advancement of stable CMS systems in pigeon pea, various CMS-based hybrids were developed and assessed (Saxena et al., 2006). Of the various CMS systems available in pigeon pea, CMS system derived from $A_2$ cytoplasm (*Cajanus scarabaeoides*) and $A_4$ cytoplasm (*Cajanus cajanifolius*) have been extensively used for breeding commercial hybrids (Choudhary and Singh 2015). Experimental testing of the newly developed CMS-hybrids was conducted by comparing them with the popular cultivars and were categorized into early, medium, and late-maturity groups. The performance of the hybrids was promising and they displayed a high degree of standard heterosis within each group. The hybrids reporting more than 40% standard heterosis were further subjected to multi-location testing (Saxena et al., 2018). Some of the CMS-based pigeon pea hybrids are briefly discussed below.

## 5.1. Early-Maturing CMS-Hybrids

In 2004, GTH 1, the world's first early maturing (140 days) CMS-based pigeon pea hybrid with $A_2$ cytoplasm was developed at SDAU (Sardarkrushinagar Dantiwada Agricultural University), Gujarat, India and released by ICAR in the state of Gujarat. GTH 1 was developed by crossing an $A_2$ CMS line GT 288A (as a female parent) with a fertility restorer GTR-11 (as a male parent). The multi-location yield testing conducted from 2000 to 2003, revealed that GTH 1 (1830 kg $ha^{-1}$) gave 57% and 32% higher yield over GMS-based hybrid AKPH 4101 and the best local cultivar GT 101, respectively. In 2003, front line demonstrations were performed in three districts and this hybrid displayed 25.3% heterosis over the popular control variety (Table 5.1). It gave the highest yield in the

Central Zone of India and was subsequently released for cultivation. However, this hybrid failed to gain its stake hold due to the instability of their fertility restorer (Saxena and Nadarajan 2010; Sameer Kumar et al., 2016; Saxena et al., 2018).

The initial panel of early-maturing hybrids with $A_4$ cytoplasm (ICPH 2433, ICPH 2438, and ICPH 2383) was developed by ICRISAT and were evaluated for in multi-location trials (Table 5.2). The data revealed that the hybrids ICPH 2433, ICPH 2438, and ICPH 2383 exhibited a high level of heterosis, with 54%, 42%, and 36% respectively over the control cultivar UPAS 120. ICPH 2433 recorded the highest yield (2419 kg ha$^{-1}$) followed by ICPH 2438 (2377 kg ha$^{-1}$) and ICPH 2429 (2164 kg ha$^{-1}$), respectively.

**Table 5.1. CMS-based hybrid GTH-1 in Front line demonstrations during *Kharif* season in the year 2003**

| **Hybrid** | **Places** | **Grain yield (Kg ha-1)** | | **% heterosis in hybrid** |
|---|---|---|---|---|
| | | Control Variety | GTH-1 | |
| CMS-based | Bolundra, | 2150 | 2800 | 30 |
| pigeon pea | Sabarkantha, Dhota, | 2210 | 2650 | 20 |
| hybrid GTH-1 | Kamalpur, Mehsana, Deodar, | 1580 | 2680 | 69 |
| | Banaskantha | 2045 | 2563 | 25 |
| | Mean | 1996 | 2673 | 33.9 |

Source: Saxena and Nadarajan 2010.

## 5.2. Medium-Maturing CMS-Hybrids

The medium-maturing group (161-200 days) holds the largest area under pigeon pea cultivation. The primary requirement for this group is resistant to diseases and in this endeavor, a CMS line ICPA 2043 was developed, and the new hybrids were developed by utilizing disease-resistant restorers (Saxena and Nadarajan 2010). This group has gained priority in breeding new hybrids and the availability of a maximum number of stable CMS and restorer lines in this group has encouraged the development of more than 3000 new hybrids (Saxena et al., 2014). These hybrids demonstrated yield superiority varying between 50 to 100% over the popular local varieties (Saxena et al., 2006). It was reported that the best hybrid ICPH 3371 (3013 kg $ha^{-1}$) exhibited a 62% yield gain. Some other promising hybrids include ICPH 3491, ICPH 3497, and ICPH 3481 which demonstrated a yield advantage of 57%, 44%, and 41%, respectively over the local varieties (Table 5.2). These new hybrids demonstrated higher resistance to wilt and sterility mosaic diseases and performed exceptionally well in different environments (Saxena et al., 2018).

**Table 5.2. Performance of CMS-based pigeon pea hybrids reported in multi-location trials across India**

| Maturity Group | CMS-based Hybrid | No. of Locations | Yield (kg $ha^{-1}$) | Standard[z] Heterosis (%) |
|---|---|---|---|---|
| Early | ICPH 2433 | 25 | 2,306** | 54 |
| | ICPH 2438 | 25 | 2,127** | 42 |
| | ICPH 2363 | 25 | 2,048** | 36 |
| Medium | ICPH 3491 | 18 | 2,919** | 57 |
| | ICPH 3497 | 18 | 2,686** | 44 |
| | ICPH 3481 | 18 | 2,637** | 41 |
| Late | ICPH 2307 | 05 | 2,855** | 53 |
| | ICPH 2306 | 05 | 2,600** | 39 |
| | ICPH 2896 | 05 | 2,579** | 38 |

Source: Saxena et al., 2018.

z – Superiority over control cultivar.

** – significantly different from the control variety at $p < 0.01\%$.

## 5.3. Late-Maturing CMS-Hybrids

The late-maturing (more than 250 days) pigeon pea group have a stringent short day photoperiod pre-requisite for the initiation of flowering. They are restricted to the regions with a day length of ≤ 10 hours. In this group, limited research has been performed concerning the exploitation of hybrid vigor. Few promising hybrids bred in this category are ICPH 2307 (53%), ICPH 2306 (39%), and ICPH 2896 (38%), determining high yield gain over the local controls (Table 5.2) (Saxena and Nadarajan 2010; Saxena et al., 2018).

## 5.4. World's First Commercial CMS-based Pigeon Pea Hybrid

ICPH 2671, was the first-ever commercial CMS-based legume hybrid, bred by crossing a male-sterile line ICPA 2043 with a restorer line ICPR267. The hybrid is medium-maturing (164-184 days), non-determinant with profuse branching. ICPH 2671 can cope with a short spells of drought owing to its higher root mass and depth. It demonstrates a high level of resistance to wilt and sterility mosaic diseases along with a high survival rate (88%) under water logging conditions (Sultana et al., 2013). During 2005-2008, this hybrid was evaluated in multi-location trials depicting average yield varying between 2022 kg $ha^{-1}$ to 2702 kg $ha^{-1}$. It demonstrated an average 35% heterosis over the local check Maruti (Table 5.3). In 2010, based on its remarkable performance, ICP 2671 was released for cultivation purposes in the state of Madhya Pradesh, India.

Following the release of ICPH 2671, two more medium duration hybrids ICPH 3762 and ICPH 2740 were released in India. In 2014, ICPH 3762 was released by the Odisha University of Agriculture and Technology, Odisha, India (Saxena et al., 2014). In multi-location trials, the hybrid recorded 20%-67% yield superiority over the popular cultivar Asha with an average yield potential of 1726 kg $ha^{-1}$. ICPH 3762 is resistant to *Fusarium* wilt and sterility mosaic disease in comparison to the local checks. It is reported to be suitable for all soil types in India (Sameer Kumar et al., 2016).

Later in 2015, ICPH 2740 was released by Professor Jayashankar Telangana State Agriculture University, Hyderabad, India. Like hybrid ICPH 3762, the hybrid ICPH 2740 out-yielded the control Asha by 42% (Saxena and Tikle 2015). These highly potential and disease-resistant pigeon pea hybrids have displayed extremely encouraging increment in the productivity levels. It is believed that the large-scale endorsement of the hybrids could break the yield plateau and will result in a boosting the national production levels of pigeon pea in India (Saxena et al., 2018).

**Table 5.3. Yield performance and standard heterosis of the CMS-based pigeon pea hybrid ICPH 2671 in multi-location testing from 2005–2008**

| Year | No. of locations | Yield (kg $ha^{-1}$) | | Standard Heterosis (%) |
|---|---|---|---|---|
| | | ICPH 2671 | Maruti variety | |
| 2005 | 5 | 3,138** | 1,855 | 69 |
| 2006 | 5 | 2,694** | 2,066 | 30 |
| 2007 | 11 | 2,702* | 2,140 | 26 |
| 2008 | 22 | 2,022* | 1,746 | 16 |
| **Mean** | | **2,639** | **1,952** | **35** |

Source: Saxena et al., 2018.

*,**– significantly different from the control at $p < 0.05$ and $p < 0.01\%$, respectively.

## 5.5. Advantages in Breeding Pigeon Pea Hybrids

The development of hybrid varieties in various crop plants like rice, maize, sorghum, sunflower, tomato, etc. has significantly revolutionized their productivity. Likewise, pigeon pea hybrids also represent various benefits over the local cultivars and these are described below:

- *Enhanced biomass and yield* - The pure line pigeon pea cultivars experience resilient competition such as weeds and inter-crops during the early stages of development, leading to poor canopy, development, and yield. On the contrary, pigeon pea hybrids exhibit more plant vigor and establish themselves more firmly in these tough situations (Saxena et al., 1992). The hybrid plants use water, sunlight, and nutrients with higher accuracy while upholding their portioning similar to the pure lines resulting in higher grain yield. ICRISAT conducted an experiment, in which 30 days old seedlings exhibited 44% and 43% greater shoot and root mass in comparison to the pure lines (Chauhan et al., 1995). The speedy growth rate of the hybrid plants supports the initiation and development of its canopy more quickly and makes them more competitive to above mentioned challenges (Saxena et al., 2006).

- *Reduced seed rates* - The outcome of the few agronomical trials conducted by ICRISAT revealed that the hybrid plants display substantial plasticity at population level varying between 16 to 66 without any harm to the seed yield (Chauhan et al., 1995). They propose that the hybrid seed rate can be cut down by a significant margin of 40-50% without losing yield per unit area. The qualities of the hybrids result in reduced hybrid seed input cost which is cost-effective to the resource-poor farmers (Saxena et al., 2006).
- *Greater disease resistance* - The two major diseases affecting pigeon pea are fusarium wilt and sterility mosaic disease which are responsible for tremendous yield loss. Few attempts have revealed that the hybrids provide greater resistance to the diseases than the pure cultivars due to their high resilience (Saxena et al., 1992). A study performed on some wilt and sterility mosaic disease resistant pigeon pea hybrids and pure cultivars in disease-free and disease-sick plots showed that the degree of disease resistance displayed by both the types was high with <1% incidence. However, under the same conditions, the hybrid vigor between the hybrids and pure lines diverges greatly. Interestingly, the hybrids displayed a 19.7% and 60% yield advantage over the pure line cultivars under disease-free and disease-sick conditions. Therefore, it was concluded that the pigeon pea hybrids possess a unique anti-fungal resistance mechanism along with added genotypic plasticity which enables them to endure and generate more yields during a stress situation than the non-hybrids (Saxena and Nadarajan 2010).
- *Greater drought tolerance* - Hybrid pigeon pea, due to its higher root mass and depth, are highly competent in drawing water from deep soil profiles. Additionally, the hybrids overcome the irregular drought situations occurring during the course of growth.

### 5.6. Constraints in Breeding Pigeon Pea Hybrids

The major constraints in pigeon pea hybrid breeding are as follows (Saxena et al., 2015):

a. Discernment of genetic variability limits the selection of heterotic hybrid parents.
b. The extensive duration of generation turnover slows down the breeding and selection process.
c. The tedious on-farm production exercise.

## 6. Hybrid Seed Production Technology

An effective approach producing quality seeds at reasonable costs is the foundation to successful hybrid breeding technology. The key determinants of a successful hybrid seed production technology are described below (Saxena et al., 2018; Saxena and Nadarajan 2010).

### 6.1. Isolation Distance

Pigeon pea exhibits a natural out-crossing (20-70%) and the extent of out-crossing is determined by the insect activity, which differs from one place to another, accordingly isolation distances should be defined for different agro-ecological areas. Based on the various reports, the isolation distance for the hybrid seed production of pigeon pea varies between 100-400 m (Saxena et al., 2006). In 2006, ICRISAT suggested an isolation distance of 500 m for producing pigeon pea hybrid seeds and until now the results are promising (Saxena 2006).

## 6.2. Isolation Plot Selection

The important factor to be considered during the selection of isolation plot is their natural habitat. The seed production area placed near the wild bushes, flowering trees, and small water bodies (natural or artificial) display perfect pod setting on the male-sterile plants. Such surroundings favorable conditions for harboring and endurance of the insects liable for cross-pollination.

## 6.3. Field Plot Technique

The adoption of a systematic field plot technique is also critical for hybrid seed production. The field layout should be designed in such a fashion that the pollens are available for longer durations, thus ensuring frequent visits of the pollinating insects, increasing the pod setting on male-sterile plants (Saxena et al., 2018). The commercial hybrid seed production technology requires large-scale production of the A (sterile)/B (maintainer) lines, restore (R) line, and hybrids (A x R). Saxena (2006) gave a detailed description of the seed production technology of the hybrids and their parents.

- *A-line:* The nucleus seed production of parent lines is significant since it affects their purity and quality. For large-scale nucleus seed production of A-line, both A-line and R-line are grown inside an insect-proof net. Both A-line and R-line plants are grown using a row ratio of 4:1 (female: male) which is results in the production of pure A-line seeds. At maturity, the crossed seed sets on the A-line plants and selfed seeds on the B-line plants are harvested separately. For producing a large-scale breeder/foundation seed of A-lines, a field plot with recommended isolation distance is chosen, A- and B-lines are planted following the advised agronomics practices.

- *B-line and R-line:* To increase seed yield of B- and R-lines, pure seed lots of both the lines are grown in separate isolations. 100 plants should be harvested from the middle section of the field plot and their progenies are grown in the subsequent generations. After assessing the purity level, the chosen progenies should be bulked to serve as the nucleus seed.
- *A x R hybrid:* The foundation seeds of A- and R-lines provides a source for the hybrid seed production. The A- and R-lines should be sown in a row ratio of 4:1 (female: male) in a separate isolated block. The row ratio should be altered depending upon the environment, insect activity and season of sowing. Added rows of pollen parent (R-line) can be grown on each side of the field plot. The visiting pollinating insects, collect the pollens from the fertile plants and performing hybridization on the male-sterile plants.

### 6.4. Seed Storage

The storage of hybrid seeds is also an important factor and requires fair protection to avert damage induced by bruchid (*Callosobruchus maculatus*) (Saxena et al., 2018). Bruchid infestation often results in compromised seed value in addition to physical injury and germination. It was observed that using Purdue Improved Cowpea storage bags remarkably minimized the bruchid destruction. It also helped in securing the germination of pigeon pea seeds (Vales et al., 2014).

## 7. Genomics Approaches for Hybrid Breeding in Pigeon Pea

To achieve success in hybrid seed technology, the process should be easy, quick, and inexpensive. Substantial advancement has been accomplished in establishing this technique in pigeon pea, however, it has few drawbacks, for instance longer generation turnover time, evaluating disease resistance, seed quality, and genetic purity. In this context, various genomic approaches have been implicated to increase the efficiency of hybrid breeding technology (Saxena et al., 2018). During the last decade, different genomic resources have been generated in pigeon pea e.g., a draft genome sequence (Singh et al., 2012; Varshney et al., 2012), mitochondrial genome sequence (Tuteja et al., 2013), chloroplast genome sequence (Kaila et al., 2016) molecular markers (Dutta et al., 2013; Saxena et al., 2014; Saxena et al., 2019), high-throughput genotyping platforms (Varshney et al., 2010), transcriptome (Saxena et al., 2020), genetic maps (Saxena et al., 2010a; Bohra et al., 2012, Arora et al., 2017) and genotyping by sequencing approach (GBS) (Saxena et al., 2011). Apart from these resources, attempts towards diversification of CMS sources in pigeon pea and understanding the molecular mechanism governing cytoplasmic male-sterility was also performed. Additionally, a different set of molecular markers were developed for fertility restoration and purity evaluation in this crop (Saxena et al., 2015). The following section describes the genomic approaches employed in the pigeon pea hybrid breeding program.

## 7.1. Molecular Mechanism of Cytoplasmic-Male Sterility in Pigeon Pea

The diversification of the stable CMS systems is an important pre-requisite for efficient pigeon pea hybrid production. Till now, nine CMS systems ($A_1$ to $A_9$) have been reported in pigeon pea, only two CMS sources $A_2$ (C. *scarabaeoides*) and $A_4$ (*C. cajanifolius*) has been commercially exploited. To gain insight on the molecular mechanism of the $A_4$ CMS system in pigeon pea, mitochondrial genome sequencing of the male-sterile (ICPA 2039) line, the maintainer (ICPB 2039) line, the hybrid (ICPH 2433) line and wild relative ICPW 29 (*C. cajanifolius*) was performed by employing Roche/454 FLX and Sangers's sequencing technology (Tuteja et al., 2013). The study, identified 13 potential *orfs* in

ICPA 2039 (male-sterile line), of which eight were located close to the mitochondrial genes and five contain portions of some other mitochondrial genes. The observations inferred that disruptions in the mitochondrial genome produced chimeric *orfs* which results in altered proteins leading to CMS.

In a subsequent study, Sinha et al., (2015) performed expression profiling and sequence variation analysis of 34 mitochondrial genes between ICPA 2039 (CMS line) and ICPA 2039 (maintainer line). They revealed a potential role of *nad4L* and *nad7* genes with the CMS for the $A_4$ cytoplasm in pigeon pea. A novel mitochondrial *orf147* was identified from the $A_4$ cytoplasm of *C. cajanifolius* which originated due to rearrangements in the CMS line (ICPA 2039). Cytotoxicity and aberrant programmed cell death were produced by *orf147* causing cytoplasmic male-sterility in pigeon pea (Bhatnagar-Mathur et al., 2018).

Recently, Saxena et al., (2020) performed comparative transcriptome analysis for exploring differentially expressed genes (DEGs) between the $A_2$ CMS system (*C. scarabaeoides*) derived male-sterile line (AKCMS11) and its fertility restorer line (AKPR303) using Illumina paired-end sequencing. They identified few candidate genes that were involved in pollen development and various metabolic processes, pointing towards their possible role in cytoplasmic male-sterility in pigeon pea. Further studies of these genes are in progress to understand their functions in conferring male-sterility in pigeon pea.

## 7.2. Genetics of Fertility Restoration

Cytoplasmic male-sterility is caused by abnormal *orfs* in the mitochondrial DNA, suppressed by the nuclear gene known as restorers of fertility (*Rf*). Therefore, screening of germplasm to identify fertility restorer (*Rf*) genes and developing stable restorers need better knowledge of the genetics of the *Rf* genes. Saxena et al., (2011) reported that the $A_4$ cytoplasm in pigeon pea was controlled by two independent dominant fertility restorer genes. The results revealed that the hybrids possessing two dominant restorer genes produced a higher number of pollens and were more stable in comparison to the ones with a single dominant restorer gene. Similarly, two dominant restorer genes were responsible for the determination of fertility restoration of the $A_2$ cytoplasm in pigeon pea (Meshram and Patil 2018). Ten markers linked with fertility restoration in pigeon pea were detected by employing linkage mapping and quantitative trait loci (QTL) analysis (Bohra et al., 2012). The utility of these markers was narrow because of low marker density on the linkage map and large genomic gaps amidst the QTL regions.

### 7.3. Hybrid Seed Purity Testing

Some important challenges in CMS-based hybrid breeding program are supplying of sufficient amount of true hybrid seeds to the breeders and sustaining the genetic purity of the parental lines of the hybrids. Any undesired impurities can significantly affect the productivity of the hybrid seeds along with the purity of the parental lines, which are crucial for the accomplishment of hybrid breeding technology (Saxena et al., 2015). To successfully develop pure CMS-based hybrids, an efficient purity estimation method is needed to evaluate the purity of the parents and affirm the hybridity. Traditionally, the 'grow-out test' (GoT) was employed for purity check of the seed samples in the majority of crops. However, GoT confides in various morphological and floral features for assessing seed purity which is time-consuming and labor-intensive. In pigeon pea, this method is not suitable because of the long generation turnover and strong photoperiod-sensitive reaction (Saxena et al., 2018). In comparison, the application of the molecular markers-based technique for purity assessment of hybrid seeds was considered a favorable option since it is cost-effective and saves time. This approach has been reported in many plant species like maize, cotton, rice, and safflower (Asif et al., 2009; Ali et al., 2008; Sundaram et al., 2008; Naresh et al., 2009).

In pigeon pea, several molecular-based markers have been developed for assessing the hybrid seed purity. The first attempt was performed by

Saxena et al., (2010b), two diagnostic nuclear SSR markers were developed in a short-duration hybrid (ICPH 2438). Bohra et al., (2012) identified a set of 42 SSR markers for purity assessment for each of the two hybrids (ICPH 2671 and ICPH 2438). Out of these, four markers for both hybrids were selected for performing multiplex assays. Subsequently, another set of seven SSR markers were detected for differentiating sterile line, maintainer line, and hybrids (Bohra et al., 2015). Sinha et al., (2015) identified the marker *nad7a_del* obtained from the *nad7* mitochondrial gene distinguishing the male-sterile line (ICPA 2039) and its corresponding maintainer line (ICPB 2039). Additionally, 24 sets of novel mitochondrial SSR markers were developed in pigeon pea that can differentiate between the CMS and their respective maintainer line (Khera et al., 2015). Bohra et al., (2017) identified ten hyper-variable SSR markers for testing three pigeon pea hybrids IPAH 16-06, IPAH 16-07 and IPH 15-03.

## 7.4. Two-Line Hybrid Breeding Approach

The three-line based hybrid breeding approach is technically challenging and costly for commercial hybrid seed production. This has led to the exploitation of a much simpler technique such as the "two-line hybrid breeding" scheme to produce hybrids. The discovery of a temperature-sensitive male-sterile (TGMS) line in pigeon pea has elucidated the approach and reduced the cost associated with hybrid breeding. A cross between *Cajanus sericeus* (wild relative) and ICPA 85010 (cultivar) produced the TGMS line (Saxena 2014). The TGMS line is evaluated under controlled environments for their male-sterility to male-fertility conversion with diverse temperature patterns. Preliminary studies on fertility transition behavior revealed that these TGMS lines respond to day temperature, transitioning to male-sterile >24°C and to male-fertile with <23°C. Also, various cytological and transcriptomic studies of the sterile and fertile anthers are being conducted to discover the potential candidate genes(s) and to gain insight into the underlying molecular basis

(Saxena et al., 2015). The detection of the putative gene(s) governing fertility reversal in this line will perform a remarkable role in breeding and developing a stable two-line hybrid system. The development of this attribute requires identification, cloning, and transferring of the key sterility gene(s) to other parental lines. Genetic study and fine mapping of TGMS trait have been already been reported in rice and wheat (Lee et al., 2005; Guo et al., 2006). Similarly, the technique is currently being undertaken by ICRISAT to discern the TGMS feature in pigeon pea (Saxena et al., 2015; Saxena et al., 2018).

## Conclusion

Pigeon pea displays great genetic variability and plasticity, however, the breeders have struggled to exploit it for enhancing productivity. In the past five decades, enormous efforts have been put in by the breeders to develop new pigeon pea short-duration varieties for improving the seed quality and overcoming the impediment of important diseases like sterility mosaic and fusarium wilt, but the national productivity has remained low and unchanged. Traditionally, manual emasculation and wind or insect pollination were used for producing hybrid seeds. Following the development of male-sterile systems, the hybrid seed production became easier and cost-effective. Subsequently, cytoplasmic male-sterile systems have emerged as a reliable tool and provided a breakthrough in enhancing the yield of many crop plants by exploiting heterosis. Pigeon pea has natural out-crossing, and the recent development of stable CMS sources along with good restorers and maintainers is a boon to pigeon pea breeders. It facilitates the production and commercialization of CMS-based hybrids and provided a platform for combating the existing yield plateau. The success was achieved by concerted efforts of ICRISAT, ICAR and SAUs through the breeding of several excellent CMS systems and the release of the world's first commercial hybrid legume. Presently, new approaches for large-scale production of hybrid seeds and corresponding parental lines have been developed by the pigeon pea breeders (Saxena et al., 2010). Till date, commercial pigeon pea hybrids have been developed from $A_2$ and $A_4$ CMS systems, but additional efforts are needed for diversification of the cytoplasm type and improvement in biotic and abiotic stress response of the parental lines. It is believed that breeding a large number of hybrids will enable easy selection of explicit hybrids displaying an increased level of hybrid vigor. The successful implementation of pigeon pea CMS systems depends on the utilization of genetically pure parental lines, expansion of hybrid vigor, efficient use of the CMS-based hybrid approach, regular practice, monitoring and assessment, and promotional or marketing exercise. Recently, the CMS-based hybrid seed technology is improved and transferred to various national agencies, public and private seed companies for developing high yielding pigeon pea hybrids. Hence, the hybrid seed technology has provided a handle for increasing pigeon pea productivity and has broken the long-awaited yield barrier in this legume crop.

## REFERENCES

Acquaah, G. 2012. "Breeding hybrid cultivars." In *Principles of Plant Genetics and Breeding* 355-373.

Ahmadikhah, A., and G. I. Karlov. (2006). "Molecular mapping of the fertility-restoration gene *Rf4* for WA-cytoplasmic male sterility in rice." *Plant Breeding* 125:363-367.

Ali, Muhammad Amjad, Muhammad Tahir Seyal, Shahid Iqbal Awan, Shahid Niaz, Shiraz Ali, and Amjad Abbas. 2008. "Hybrid authentication in upland cotton through RAPD analysis." *Australian Journal of Crop Science* 2:141-149.

Ariyanayagam, R. P., A. Nageshwar Rao, and P. P. Zaveri. 1993. "Gene-cytoplasmic male-sterility in pigeonpea." *International Pigeonpea Newsletter* 18:7-11.

Ariyanayagam, R. P., A. Nageshwar Rao, and P. P. Zaveri. 1995. "Cytoplasmic-Genic Male-Sterility in Interspecific Matings of *Cajanus*." *Crop Science* 35:981-985.

Arora, Sheetal, Ajay Kumar Mahato, Sangeeta Singh, Paritra Mandal, Shefali Bhutani, Sutapa Dutta, Giriraj Kumawat et al., 2017. “A high-density intra-specific SNP linkage map of pigeonpea (*Cajanas cajan* L. Millsp.).” *PloS one* 12:e0179747.

Asif, Muhammad, M. Ur Rahman, Javed Iqbal Mirza, and Yusuf Zafar. 2009. “Parentage confirmation of cotton hybrids using molecular markers.” *Paksitan Journal of Botany* 41:695-701.

Bailey-Serres, Julia, Deborah K. Hanson, Thomas D. Fox, and Christopher J. Leaver. 1986. “Mitochondrial genome rearrangement leads to extension and relocation of the cytochrome c oxidase subunit I gene in sorghum.” *Cell* 47:567-576.

Barclay, Adam. 2010 “Hybridizing the world.” *Rice Today* 9:32-35.

Bateson, William, and A. E. Gairdner. 1921 “Male-sterility in flax, subject to two types of segregation.” *Journal of Genetics* 11:269-275.

Bentolila, Stéphane, Antonio A. Alfonso, and Maureen R. Hanson. 2002. “A pentatricopeptide repeat-containing gene restores fertility to cytoplasmic male-sterile plants.” *Proceedings of the National Academy of Sciences* 99:10887-10892.

Bhatnagar-Mathur, Pooja, Ranadheer Gupta, Palakolanu Sudhakar Reddy, Bommineni Pradeep Reddy, Dumbala Srinivas Reddy, C. V. Sameerkumar, Rachit Kumar Saxena, and Kiran K. Sharma. 2018. “A novel mitochondrial *orf147* causes cytoplasmic male sterility in pigeonpea by modulating aberrant anther dehiscence.” *Plant Molecular Biology* 97:131-147.

Bohra, Abhishek, I. P. Singh, Ashutosh K. Yadav, Abhinav Pathak, K. R. Soren, S. K. Chaturvedi, and N. Nadarajan. 2015. “Utility of informative SSR markers in the molecular characterization of cytoplasmic genetic male sterility-based hybrid and its parents in pigeonpea.” *National Academy Science Letters* 38:13-19.

Bohra, Abhishek, Rachit K. Saxena, B. N. Gnanesh, Kulbhushan Saxena, M. Byregowda, Abhishek Rathore, P. B. KaviKishor, Douglas R. Cook, and Rajeev K. Varshney. 2012. “An intra-specific consensus genetic map of pigeonpea [*Cajanus cajan* (L.) Millspaugh] derived from six mapping populations." *Theoretical and Applied Genetics* 125:1325-1338.

Bohra, Abhishek, Rintu Jha, Gaurav Pandey, Prakash G. Patil, Rachit K. Saxena, Indra P. Singh, D. Singh et al., 2017. "New hypervariable SSR markers for diversity analysis, hybrid purity testing and trait mapping in Pigeonpea [*Cajanus cajan* (L.) Millspaugh]." *Frontiers in Plant Science* 8:377.

Bonhomme, Sandrine, Françoise Budar, Dominique Lancelin, Ian Small, Marie-Christine Defrance, and Georges Pelletier. 1992. "Sequence and transcript analysis of the *Nco2. 5*Ogura-specific fragment correlated with cytoplasmic male sterility in *Brassica*cybrids." *Molecular and General Genetics* 235:340-348.

Budar, Françoise, and Georges Pelletier. 2001. "Male sterility in plants: occurrence, determinism, significance and use." *Comptes Rendus de l'Académie des Sciences-Series III-Sciences de la Vie* 324:543-550.

Budar, Françoise, and Richard Berthomé. 2007. "Cytoplasmic male sterilities and mitochondrial gene mutations in plants." In *Plant mitochondria: annual plant reviews* edited by D.C Logan, 278-307. Blackwell Publishing Ltd., Oxford, UK.

Chase, Christine D. 2007 "Cytoplasmic male sterility:a window to the world of plant mitochondrial–nuclear interactions." *Trends in Genetics* 23:81-90.

Chase, Christine D., and S. Gabay-Laughnan. 2004. "Cytoplasmic male sterility and fertility restoration by nuclear genes." In *Molecular biology and biotechnology of plant organelles*, edited by Henry Daniell and Christine Chase, 593-622. Springer, Dordrecht.

Chaudhari, Sunil, A. N. Tikle, Uttamchand, K. B. Saxena, and A. Rathore. 2015. "Stability of Cytoplasmic Genetic Male Sterility and Fertility Restoration in Pigeonpea." *Journal of Crop Improvement* 29:269-280.

Chauhan, Y. S., C. Johansen, and K. B. Saxena. 1995. "Physiological Basis of Yield Variation in Shortduration Pigeonpea Grown in Different Environments of the Semi-Arid Tropics." *Journal of Agronomy and Crop Science* 174:163-171.

Chen, Letian, and Yao-Guang Liu. 2014. "Male sterility and fertility restoration in crops." *Annual review of Plant Biology* 65:579-606.

Chittenden, R. J., and Caroline Pellew. 1927. "A suggested interpretation of certain cases of anisogeny." *Nature* 119:10-11.

Choudhary, Arbind K. and Indra Prakash Singh. 2015. "A study on comparative fertility restoration in $A_2$ and $A_4$ cytoplasms and its implication in breeding hybrid pigeonpea [*Cajanus cajan* (L.) Millspaugh]." *American Journal of Plant Sciences* 6:385-391.

Correns, C. 1908. "Die poller dev mannlichen keimzellen bei de. Geschle chtsbestimmung dev gynodiocischen Pflanzen." *Berichte der Deutschen Botanischen Gesellschaft*, 26:686-701.

Damaris, A. Odeny. 2007. "The potential of pigeonpea (*Cajanus cajan* (L.) Millsp.) in Africa." *Natural Resources Forum* 31:297-305.

Das, Pritam, Biswarup Mukherjee, Chand Kumar Santra, Suparna Gupta, and Tapash Dasgupta. 2014. "Agro-botanical Characterization of some released $F_1$ hybrids in rice (*Oryza sativa*L.)." *International Journal of Science Research* 4:1-9.

Das, Srirupa, Supriya Sen, Anirban Chakraborty, Papia Chakraborti, Mrinal K. Maiti, Asitava Basu, Debabrata Basu, and Soumitra K. Sen. 2010. "An unedited 1.1 kb mitochondrial *orfB* gene transcript in the wild abortive cytoplasmic male sterility (WA-CMS) system of *Oryza sativa* L. subsp. *indica*." *BMC Plant Biology* 10:39.

De, D. N. 1974. "Pigeonpea." In *Evolutionary studies on world crops*, Diversity and change in the Indian Subcontinent, edited by J. Hutchinson, 79-87. Cambridge University Press, Cambridge.

Deshmukh, N. Y. 1959. "Sterile mutants in tur (*Cajanus cajan*)." *Nagpur Agriculture College Marg* 33:20-21.

Dewey, R. E., D. H. Timothy, and C. S. Levings. 1987. "A mitochondrial protein associated with cytoplasmic male sterility in the T cytoplasm of maize." *Proceedings of the National Academy of Sciences USA* 84:5374-5378.

Dewey, R. E., D. H. Timothy, and C. S. Levings. 1991. "Chimeric mitochondrial genes expressed in the C male-sterile cytoplasm of maize." *Current Genetics* 20:475-482.

Dohmen, Gabriel, and Paul Tudzynski. 1994. "A DNA-polymerase-related reading frame *(pol-r)* in the mtDNA of *Secale cereale*." *Current Genetics* 25:59-65.

Ducos, Eric, Pascal Touzet, and Marc Boutry. 2001. "The male sterile G cytoplasm of wild beet displays modified mitochondrial respiratory complexes." *The Plant Journal* 26:171-180.

Dudhe, M. Y., M. K. Moon, and S. S. Lande. 2011. "Study of gene action for restorer lines in sunflower." *Helia* 34:159-164.

Dundas, Ian S, Kul B. Saxena, and D. E. Byth. "Pollen mother cell and anther wall development in a photoperiod-insensitive male-sterile mutant of pigeon pea (*Cajanus cajan* (L.) Millsp.) "*Euphytica* 31:371-375.

Dutta, Sutapa, Ajay K. Mahato, Priti Sharma, Ranjeet S. Raje, Tilak R. Sharma, and Nagendra K. Singh. 2013. "Highly variable 'Arhar' simple sequence repeat markers for molecular diversity and phylogenetic studies in pigeonpea [*Cajanus cajan* (L.) Millisp.]." *Plant Breeding* 132:191-196.

Duvick, Donald N. 1959. "The use of cytoplasmic male-sterility in hybrid seed production." *Economic Botany* 13:167-195.

Eckardt, Nancy A. 2006. "Cytoplasmic male sterility and fertility restoration." *Plant Cell*, 18:515–517.

FAO, 2018. *FAOSTAT*. Updated on Nov 24, 2020. http://www.fao.org/faostat/en/#home.

Forde, Brian G., Robin JC Oliver, and Christopher J. Leaver. 1978. "Variation in mitochondrial translation products associated with male-sterile cytoplasms in maize." *Proceedings of the National Academy of Sciences USA* 75:3841-3845.

Fujii, Sota, and Kinya Toriyama. 2009. "Suppressed expression of Retrograde-Regulated Male Sterility restores pollen fertility in cytoplasmic male sterile rice plants." *Proceedings of the National Academy of Sciences USA* 106:9513-9518.

Fujii, Sota, Setsuko Komatsu, and Kinya Toriyama. 2007. "Retrograde regulation of nuclear gene expression in CW-CMS of rice." *Plant Molecular Biology* 63:405-417.

Fujii, Sota, Tomohiko Kazama, Mari Yamada, and Kinya Toriyama. 2010. "Discovery of global genomic re-organization based on comparison of two newly sequenced rice mitochondrial genomes with cytoplasmic male sterility-related genes." *BMC Genomics* 11:1-15.

Geiger, H. H., and F. W. Schnell. 1970. "Cytoplasmic male sterility in Rye (*Secale cereale*L.)." *Crop science* 10:590-593.

Godfray, H. Charles J., John R. Beddington, Ian R. Crute, Lawrence Haddad, David Lawrence, James F. Muir, Jules Pretty, et al., 2010. "Food security: the challenge of feeding 9 billion people." *Science* 327:812-818.

Gulyas, Gergely, Youngsup Shin, Hoytaek Kim, Jang-Soo Lee, and Yutaka Hirata. 2010. "Altered transcript reveals an *orf507* sterility-related gene in chili pepper (*Capsicum annuum* L.)." *Plant Molecular Biology Reporter* 28:605-612.

Guo, R. X., D. F. Sun, Z. B. Tan, D. F. Rong, and C. D. Li. 2006. "Two recessive genes controlling thermophotoperiod-sensitive male sterility in wheat." *Theoretical and Applied Genetics* 112:1271-1276.

Gupta, S. C., and D. G. Faris. 1983. "New steriles in early pigeonpea." *International Pigeonpea News letter* 2:21-22.

Hagihara, E., N. Itchoda, Y. Habu, S. Iida, T. Mikami, and T. Kubo. 2005. "Molecular mapping of a fertility restorer gene for Owen cytoplasmic male sterility in sugar beet." *Theoretical and Applied Genetics* 111:250-255.

Hanson, Maureen R. 1991. "Plant mitochondrial mutations and male sterility." *Annual Review of Genetics* 25:461-486.

Hanson, Maureen R., and Stéphane Bentolila. 2004. "Interactions of mitochondrial and nuclear genes that affect male gametophyte development." *The Plant Cell* 16:S154-S169.

Hariprasanna, K., and J. V. Patil. 2015. "Sorghum: origin, classification, biology and improvement." In *Sorghum molecular breeding*, edited by Madhusudhana R, Rajendra Kumar P. and J. V. Patil, 3-220, Springer, New Delhi.

Havey, Michael J. 2004. "The use of cytoplasmic male sterility for hybrid seed production." In *Molecular biology and biotechnology of plant organelles*, edited by Henry Daniell and Christine Chase, 623-634. Springer, Dordrecht.

Horn, R., R. H. Köhler, A. Loessl, R. Kräuter, J. R. Gerlach, J. E. G. Hustedt, V. Hahn et al., 1995. "Development and molecular analysis of alloplasmatic male-sterility in sunflower." In *Advances in Plant Breeding* 18, *Genetic Mechanisms for Hybrid Breeding*, edited by U. Kück and G. Wricke, 89-110. Blackwell Wissenschafts-Verlag.

Horn, Renate, Joachim EG Hustedt, Andreas Horstmeyer, Josef Hahnen, Klaus Zetsche, and Wolfgang Friedt. 1996. "The CMS-associated 16 kDa protein encoded by *orfH522* in the PET1 cytoplasm is also present in other male-sterile cytoplasms of sunflower." *Plant Molecular Biology* 30:523-538.

Howad, Werner, and Frank Kempken. 1997. "Cell type-specific loss of atp6 RNA editing in cytoplasmic male sterile *Sorghum bicolor*." *Proceedings of the National Academy of Sciences* 94:11090-11095.

Howard, Albert, Gabrielle LC Howard, and Rahman Abdur. 1919. "Studies in the pollination of Indian crops." *Memoirs, Department of Agriculture India Botanical Series* 10:195-200.

Hu, Jun, Kun Wang, Wenchao Huang, Gai Liu, Ya Gao, Jianming Wang, Qi Huang et al., 2012. "The rice pentatricopeptide repeat protein RF5 restores fertility in Hong-Lian cytoplasmic male-sterile lines via a complex with the glycine-rich protein GRP162." *The Plant Cell* 24:109-122.

Igarashi, Keisuke, Tomohiko Kazama, Keiji Motomura, and Kinya Toriyama. 2013. "Whole genomic sequencing of RT98 mitochondria derived from Oryza rufipogon and northern blot analysis to uncover a cytoplasmic male sterility-associated gene." *Plant and Cell Physiology* 54:237-243.

Ingale, B. V., N. D. Jambhale, B. D. Waghmode, and V. V. Dalvi. 2006. "Sahyadri 2, an early rice hybrid for Maharashtra State in India." *International Rice Research Notes* 31:20-21.

Itabashi, Etsuko, Natsuko Iwata, Sota Fujii, Tomohiko Kazama, and Kinya Toriyama. 2011. "The fertility restorer gene, *Rf2*, for Lead Rice-type cytoplasmic male sterility of rice encodes a mitochondrial glycine-rich protein." *The Plant Journal* 65:359-367.

Itabashi, Etsuko, Tomohiko Kazama, and Kinya Toriyama. 2009. "Characterization of cytoplasmic male sterility of rice with Lead Rice cytoplasm in comparison with that with Chinsurah Boro II cytoplasm." *Plant Cell Reports* 28:233-239.

Iwabuchi, Mari, Junko Kyozuka, and K. O. Shimamoto. 1993. "Processing followed by complete editing of an altered mitochondrial *atp6* RNA restores fertility of cytoplasmic male sterile rice." *The EMBO Journal* 12:1437-1446.

Iwabuchi, Mari, Nobuya Koizuka, Hideya Fujimoto, Takako Sakai, and Jun Imamura. 1999. "Identification and expression of the kosena radish (*Raphanus sativus* cv. Kosena) homologue of the ogura radish CMS-associated gene, *orf138*." *Plant Molecular Biology* 39:183-188.

Ji, Jiaojiao, Wei Huang, Chuanchuan Yin, and Zhenhui Gong. 2013. "Mitochondrial cytochrome c oxidase and $F_1F_o$-ATPase dysfunction in peppers (*Capsicum annuum* L.) with cytoplasmic male sterility and its association with *orf507* and *Ψatp6*-2 genes." *International journal of Molecular Sciences* 14:1050-1068.

Johns, Carol, Meiqing Lu, Anna Lyznik, and Sally Mackenzie. 1992. "A mitochondrial DNA sequence is associated with abnormal pollen development in cytoplasmic male sterile bean plants." *The Plant Cell* 4:435-449.

Jones, Henry A. and A. E. Clarke. 1943. "Inheritance of male sterility in the onion and the production of hybrid seed." *Proceedings for the American Society for Horticultural Science* 43:189-194.

Kaila, Tanvi, Pavan K. Chaduvla, Swati Saxena, Kaushlendra Bahadur, Santosh J. Gahukar, Ashok Chaudhury, T. R. Sharma, N. K. Singh, and Kishor Gaikwad. 2016. "Chloroplast genome sequence of Pigeonpea (*Cajanus cajan* (L.) Millspaugh) and *Cajanus scarabaeoides* (L.) Thouars: Genome organization and comparison with other legumes." *Frontiers in Plant Science* 7:1847.

Kaul, M. L. K. 1988. "Male sterility in higher plants. Grossman." *Monographs on Theoretical and Applied Genetics* 10:15-95.

Kazama, Tomohiko, Takahiro Nakamura, Masao Watanabe, Mamoru Sugita, and Kinya Toriyama. 2008. "Suppression mechanism of mitochondrial ORF79 accumulation by Rf1 protein in BT-type cytoplasmic male sterile rice." *The Plant Journal* 55:619-628.

Khera, Pawan, Rachit Saxena, C. V. Sameerkumar, Kulbhushan Saxena, and Rajeev K. Varshney. 2015. "Mitochondrial SSRs and their utility in distinguishing wild species, CMS lines and maintainer lines in pigeonpea (*Cajanus cajan* L.)." *Euphytica* 206:737-746.

Kim, Dong Hwan, Jeong Gu Kang, and Byung-Dong Kim. 2007. "Isolation and characterization of the cytoplasmic male sterility-associated *orf456* gene of chili pepper (*Capsicum annuum* L.)." *Plant Molecular Biology* 63:519-532.

Kim, Yu-Jin, and Dabing Zhang. 2018. "Molecular control of male fertility for crop hybrid breeding." *Trends in plant science* 23:53-65.

Köhler, Rainer Hans, Renate Horn, Andreas Lössl, and Klaus Zetsche. 1991. "Cytoplasmic male sterility in sunflower is correlated with the co-transcription of a new open reading frame with the *atpA* gene." *Molecular and General Genetics* 227:369-376.

Koizuka, Nobuya, Ritsuko Imai, Hideya Fujimoto, Takahiko Hayakawa, Yusuke Kimura, Junko Kohno-Murase, Takako Sakai et al., 2003. "Genetic characterization of a pentatricopeptide repeat protein gene, orf687, that restores fertility in the cytoplasmic male-sterile Kosena radish." *The Plant Journal* 34:407-415.

Kölreuter, Joseph Gottlieb. 1763. *Vorläufige Nachricht von einigen das Geschlecht der Pflanzen betreffenden Versuchen und Beobachtungen: Fortsetzung*. Gleditsch.

Landgren, Maria, Mattias Zetterstrand, Eva Sundberg, and Kristina Glimelius. 1996. "Alloplasmic male-sterile Brassica lines containing *B. tournefortii* mitochondria express an ORF 3′ of the *atp6* gene and a 32 kDa protein." *Plant Molecular Biology* 32:879-890.

Laser, Kenneth D., and Nels R. Lersten. 1972. "Anatomy and cytology of microsporogenesis in cytoplasmic male sterile angiosperms." *The Botanical Review* 38:425-454.

Laver, H. K., S. J. Reynolds, F. Moneger, and C. J. Leaver. 1991. "Mitochondrial genome organization and expression associated with cytoplasmic male sterility in sunflower (*Helianthus annuus*)." *The Plant Journal* 1:185-193.

Leclercq, P. 1969. "Une stérilité male cytoplasmic chez le tournesol. Ann." *Amélior Plant* 19:99-106. ["Cytoplasmic male sterility in sunflower. Ann. " *Improved Plant*]

Lee, Dong Sun, Li Juan Chen, and Hak Soo Suh. 2005. "Genetic characterization and fine mapping of a novel thermo-sensitive genic male-sterile gene *tms6* in rice (*Oryza sativa* L.)." *Theoretical and Applied Genetics* 111:1271-1277.

Levings 3rd, C. S. 1993. "Thoughts on cytoplasmic male sterility in cms-T maize." *The Plant Cell* 5:1285-1290.

Levings III, C. S., and R. E. Dewey. 1988. "Molecular studies of cytoplasmic male sterility in maize." *Philosophical Transactions of the Royal Society of London. B, Biological Sciences* 319:177-185.

Levings, Charles S. 1990. "The Texas cytoplasm of maize: cytoplasmic male sterility and disease susceptibility." *Science* 250:942-947.

L'Homme, Yvan, Richard J. Stahl, Xiu-Qing Li, Amina Hameed, and Gregory G. Brown. 1997. "Brassica nap cytoplasmic male sterility is associated with expression of a mtDNA region containing a chimeric gene similar to the pol CMS-associated orf224 gene." *Current Genetics* 31:325-335.

Liu, H. L., T. D. Fu, and S. N. Yang. 1988. "Discovery and studies on Polima CMS line." In *7th International Rapeseed Congress/convened under the patronage of Stanislaw Zieba; by the Plant Breeding and Acclimatization Institute under the auspices of the Group Consultatif International de Recherche sur le Colza*. 69-78, 11-14 May, Poznan: Panstwowe Wydawnictwo Rolnicze i Lesne.

Liu, Zhen-Lan, Hong Xu, Jing-Xin Guo, and Yao-Guang Liu. 2007. "Structural and expressional variations of the mitochondrial genome conferring the wild abortive type of cytoplasmic male sterility in rice." *Journal of Integrative Plant Biology* 49:908-914.

Mackenzie, Sally A. 2005. "The influence of mitochondrial genetics on crop breeding strategies." *Plant Breeding Reviews* 25:115-138.

Makaroff, Christopher A., Ingrid J. Apel, and Jeffrey D. Palmer. 1989."The atp6 coding region has been disrupted and a novel reading frame generated in the mitochondrial genome of cytoplasmic male-sterile radish." *Journal of Biological Chemistry* 264:11706-11713.

Mallikarjuna, Nalini, and J. P. Moss. 1995. "Production of hybrids between *Cajanus platycarpus*and *Cajanus cajan*." *Euphytica* 83:43-46.

Mallikarjuna, Nalini, and K. B. Saxena. 2002. "Production of hybrids between *Cajanus acutifolius* and *C. cajan*." *Euphytica* 124:107-110.

Mallikarjuna, Nalini, Deepak Jadhav, and Prabhakar Reddy. 2006. "Introgression of *Cajanus platycarpus* genome into cultivated pigeonpea, *C. cajan*." *Euphytica* 149:161-167.

Mallikarjuna, Nalini, Deepak R. Jadhav, Sandhya Srikanth, and Kulbhushan B. Saxena. 2011. "*Cajanus platycarpus* (Benth.) Maesen as the donor of new pigeonpea cytoplasmic male sterile (CMS) system." *Euphytica* 182:65-71.

Mallikarjuna, Nalini, Kulbushan Saxena, Jhansi Lakshmi, Rajeev Varshney, Sandhya Srikanth, and Deepak Jadhav. 2012. "Differences between *Cajanus cajan* (L.) Millspaugh and *C. cajanifolius* (Haines) van der Maesen, the progenitor species of pigeonpea." *Genetic resources and crop evolution* 59:411-417.

Matsuhira, Hiroaki, Hiroyo Kagami, Masayuki Kurata, Kazuyoshi Kitazaki, Muneyuki Matsunaga, Yuko Hamaguchi, Eiki Hagihara et al., 2012. "Unusual and typical features of a novel restorer-of-fertility gene of sugar beet (*Beta vulgaris* L.)." *Genetics* 192:1347-1358.

McDermott, Paul, Vincent Connolly, and Tony A. Kavanagh. 2008. "The mitochondrial genome of a cytoplasmic male sterile line of perennial ryegrass (*Lolium perenne* L.) contains an integrated linear plasmid-like element." *Theoretical and Applied Genetics* 117:459-470.

Meshram, M. P., and A. N. Patil. 2018. "Genetics of fertility restoration in $A_2$ cytoplasm based hybrids of pigeonpea." *International Journal of Current Microbiology and Applied Science* 6:565-571.

Murugarajendran C., W M. Ali Khan, R Rathnaswamy, S R Sree Rangasamy, R Ravikesavan, T Kalaimagal, Kul B. Saxena and R Vijay Kumar. 1995. "IPH 732 – a new hybrid pigeonpea for Tamil Nadu". *International Chickpea and Pigeonpea Newsletter* 2:55-57.

Nakajima, Yuki, Toshiya Yamamoto, Toshiya Muranaka, and Kenji Oeda. 2001. "A novel orfB-related gene of carrot mitochondrial genomes that is associated with homeotic cytoplasmic male sterility (CMS)." *Plant Molecular Biology* 46:99-107.

Naresh, V., K. N. Yamini, P. Rajendrakumar, and V. Dinesh Kumar. 2009. "EST-SSR marker-based assay for the genetic purity assessment of safflower hybrids." *Euphytica* 170:347-353.

Niranjan, S. K. D. F. F., M. C. S. Bantilan, P. K. Joshi, and K. B. Saxena. 1998. "Evaluating hybrid technology for pigeonpea." In *Assessing joint research impacts: Proceedings of an International Workshop on Joint Impact Assessment of NARS/ICRISAT Technologies for the Semi-Arid Tropics, 2-4 Dec 1996, ICRISAT, Patancheru, India*, vol. 502:231-240.

Okazaki, Masayuki, Tomohiko Kazama, Hayato Murata, Keiji Motomura, and Kinya Toriyama. 2013. "Whole mitochondrial genome sequencing and transcriptional analysis to uncover an RT102-type cytoplasmic male sterility-associated candidate gene derived from *Oryza rufipogon*." *Plant and Cell Physiology* 54:1560-1568.

Okuda, Kenji, Fumiyoshi Myouga, Reiko Motohashi, Kazuo Shinozaki, and Toshiharu Shikanai. 2007. "Conserved domain structure of pentatricopeptide repeat proteins involved in chloroplast RNA editing." *Proceedings of the National Academy of Sciences USA* 104:8178-8183.

Pandey, N., C. B. Ojha, P. N. Narula, and S. K. Chowdhury. "1994. Bi and tri carpels and male-sterility in pigeonpea." *Indian Journal of Pulses Research* 7:62.

Park, Jee Young, Young-Pyo Lee, Jonghoon Lee, Beom-Soon Choi, Sunggil Kim, and Tae-Jin Yang. 2013. "Complete mitochondrial genome sequence and identification of a candidate gene responsible for cytoplasmic male sterility in radish (*Raphanus sativus* L.) containing DCGMS cytoplasm." *Theoretical and Applied Genetics* 126:1763-1774.

Priyadarshan, P. M. 2019. "Male Sterility." In *Plant Breeding: Classical to Modern*, 105-129. Springer, Singapore.

Rajeshwari, R., S. Sivaramakrishnan, R. L. Smith, and N. C. Subrahmanyam. 1994. "RFLP analysis of mitochondrial DNA from cytoplasmic male-sterile lines of pearl millet." *Theoretical and Applied Genetics* 88:441-448.

Rathnaswamy, R., J. L. Yolanda, T. Kalaimagal, M. Suryakumar, and D. Sassikumar. 1999. "Cytoplasmic-genic male-sterility in pigeonpea (*Cajanus cajan*)." *Indian Journal of Agricultural Science* 69:159-160.

Ratnaparkhe, M. B., V. S. Gupta, MR Ven Murthy, and P. K. Ranjekar. 1995. "Genetic fingerprinting of pigeonpea [*Cajanus cajan* (L.) Millsp.] and its wild relatives using RAPD markers." *Theoretical and Applied Genetics* 91:893-898.

Reddy, Belum VS, John M. Green, and S. S. Bisen. 1978. "Genetic Male Sterility in Pigeon Pea." *Crop Science* 18:362-364.

Reddy, L. J., and D. G. Faris. 1981. "A cytoplasmic-genetic male sterile line in pigeonpea." *International Pigeonpea Newsletter* 1:16-17.

Rhoades, Marcus M. 1931. "Cytoplasmic inheritance of male sterility in *Zea mays*." *Science* 73:340-341.

Rhoades, Marcus M. 1933. "The cytoplasmic inheritance of male sterility in *Zea mays*." *Journal of Genetics* 27:71- 95.

Rogers, John S., and John R. Edwardson. "The utilization of cytoplasmic male-sterile inbreds in the production of corn hybrids 1." *Agronomy Journal* 44:8-13.

Rouwendal, G. J. A., J. Creemers-Molenaar, and F. A. Krens. 1992. "Molecular aspects of cytoplasmic male sterility in perennial ryegrass (*Lolium perenne* L.): mtDNA and RNA differences between plants with male-sterile and fertile cytoplasm and restriction mapping of their *atp6* and *coxI* homologous regions." *Theoretical and Applied Genetics* 83:330-336.

Sabar, Mohammed, Rosine De Paepe, and Yaroslav de Kouchkovsky. 2000. "Complex I impairment, respiratory compensations, and photosynthetic decrease in nuclear and mitochondrial male sterile mutants of *Nicotiana sylvestris*." *Plant Physiology* 124:1239-1250.

Sameer Kumar, C. V., SUHAS P. Wani, Nagesh Kumar, P. Jaganmohan Rao, K. B. Saxena, A. J. Hingane, C. Sudhakar et al., 2016. "Hybrid Technology–a new vista in pigeonpea breeding." *The Journal of Research PJTSAU* 44:1-13.

Sawargaonkar, S. L., I. A. Madrap, and K. B. Saxena. 2012. "Stability of cytoplasmic male-sterile lines in pigeonpea under different month temperature." *Green farming* 3:515-517.

Saxena, K. B. 2008. "Genetic improvement of pigeon pea—a review." *Tropical Plant Biology* 1:159-178.

Saxena, K. B. 2014. "Temperature-sensitive male-sterility system in pigeonpea." *Current Science* 107:277-281.

Saxena, K. B. 2006. *Seed Production Systems in Pigeonpea ICRISAT*. International Crops Research Institute for the Semi-Arid Tropics, Patancheru, Andhra Pradesh, India.

Saxena, K. B., and A. N. Tikle. 2015. "Believe it or not, hybrid technology is the only way to enhance pigeonpea yields." *International Journal of Scientific and Research* 5:1-7.

Saxena, K. B., and N. Nadarajan. 2010. "Prospects of pigeonpea hybrids in Indian agriculture." *Electronic Journal of Plant Breeding* 1:1107-1117.

Saxena, K. B., and R. V. Kumar. 2001. "Genetics of a new male-sterility locus in pigeonpea (*Cajanus cajan* [L.] Millsp.)." *Journal of Heredity* 92:437-439.

Saxena, K. B., and R. V. Kumar. 2003. "Development of a cytoplasmic nuclear male-sterility system in pigeonpea using *C. scarabaeoides* (L.) Thouars." *Indian Journal of Genetics and Plant Breeding* 63:225-229.

Saxena, K. B., C. V. Sameerkumar, M. G. Mula, R. V. Kumar, S. B. Patil, M. Sharma, R. K. Saxena, and R. K. Varshney. 2014. "Pigeonpea hybrid ICPH 3762 (Parbati)." *Release proposal submission of crop variety to Odisha State Seed Sub-Committee, Directorate of Research, Orissa University of Agriculture & Technology, Bhubaneswar* 751, no. 003.

Saxena, K. B., D. Sharma, and M. I. Vales. 2018. "Development and commercialization of CMS pigeonpea hybrids." In *Plant Breeding Reviews*, edited by Irwin Goldman, 103-167. John Wiley & Sons, Inc.

Saxena, K. B., E. S. Wallis, and D. E. Byth. 1983. "A new gene for male sterility in pigeonpea (*Cajanus cajan* (L). Millsp.)." *Heredity* 51:419-421.

Saxena, K. B., Laxman Singh, and M. D. Gupta. 1990. "Variation for natural out-crossing in pigeonpea." *Euphytica* 46:143-148.

Saxena, K. B., Laxman Singh, R. V. Kumar, and A. N. Rao. 1996. "Development of cytoplasmic-genic male-sterility (CMS) system in pigeonpea at ICRISAT Asia Center." *Proceedings of the Working Group on Cytoplasmic-Genic Male-Sterility (CMS) in Pigeonpea*:9-10.

Saxena, K. B., R. Sultana, N. Mallikarjuna, R. K. Saxena, R. V. Kumar, S. L. Sawargaonkar, and R. K. Varshney. 2010. "Male-sterility systems in pigeonpea and their role in enhancing yield." *Plant Breeding* 129:125-134.

Saxena, K. B., R. Sultana, R. K. Saxena, R. V. Kumar, J. S. Sandhu, A. Rathore, PB Kavi Kishor, and R. K. Varshney. 2011. "Genetics of fertility restoration in $A_4$-based, diverse maturing hybrids of pigeonpea [*Cajanus cajan* (L.) Millsp.]." *Crop Science* 51:574-578.

Saxena, K. B., R. V. Kumar, and P. V. Rao. 2002. "Pigeonpea nutrition and its improvement." *Journal of Crop production* 5:227-260.

Saxena, K. B., R. V. Kumar, K. Madhavi Latha, and V. A. Dalvi. 2006. "Commercial pigeonpea hybrids are just a few steps away." *Indian Journal of Pulses Research* 19:7-16.

Saxena, K. B., R. V. Kumar, Namita Srivastava, and Bao Shiying. 2005. "A cytoplasmic-nuclear male-sterility system derived from a cross between *Cajanus cajanifolius* and *Cajanus cajan*." *Euphytica* 145:289-294.

Saxena, K. B., D. E. Byth, I. S. Dundas, and E. S. Wallis. 1982. "Genetic control of sparse pollen production in pigeon pea." *International Pigeonpea Newsletter*: 17-18.

Saxena, K. B., Y. S. Chauhan, C. Johansen, and Laxman Singh. 1992. "Recent developments in hybrid pigeonpea research." *In New Frontiers in Pulses Research and Development*, edited by Banpot Napompeth and S. Subhadra Bandhu, 58-69. Directorate of Pulses Research, Kanpur, India.

Saxena, Kul Bhushan, Ravikoti Vijaya Kumar, Ashok Narayanrao Tikle, Mukesh Kumar Saxena, Virendra Singh Gautam, Surapaneni Koteshwar Rao, Dhirendra Kumar Khare et al., 2013. "ICPH 2671–the world's first commercial food legume hybrid." *Plant Breeding* 132:479-485.

Saxena, Kul Bhushan. 2002. "Heterosis breeding in pulses-problems and prospects." 61-72. *Paper presented in Pulses for Sustainable Agriculture and Nutritional Security Proceedings of National Symposium*, New Delhi, India, April 17-19, 2001.

Saxena, M. K., Usha Saxena, K. B. Saxena, V. S. Khandalkar, and Rafat Sultana. 2011. "Profitability and production cost of hybrid pigeonpea seed." *Electronic Journal of Plant Breeding* 2:409-412.

Saxena, R. K., K. B. Saxena, R. V. Kumar, D. A. Hoisington, and R. K. Varshney. 2010a. "Simple sequence repeat-based diversity in elite pigeonpea genotypes for developing mapping populations to map resistance to Fusarium wilt and sterility mosaic disease." *Plant Breeding* 129:135-141.

Saxena, Rachit K., Kulbhushan Saxena, and Rajeev K. Varshney. 2010b. "Application of SSR markers for molecular characterization of hybrid parents and purity assessment of ICPH 2438 hybrid of pigeonpea [*Cajanus cajan* (L.) Millspaugh]." *Molecular Breeding* 26:371-380.

Saxena, Rachit K., Eric Von Wettberg, Hari D. Upadhyaya, Vanessa Sanchez, Serah Songok, Kulbhushan Saxena, Paul Kimurto, and Rajeev K. Varshney. 2014. "Genetic diversity and demographic history of *Cajanus* spp. illustrated from genome-wide SNPs." *PLoS One* 9:e88568.

Saxena, Rachit K., K. B. Saxena, Lekha T. Pazhamala, Kishan Patel, Swathi Parupalli, C. V. Sameerkumar, and Rajeev K. Varshney. 2015. "Genomics for greater efficiency in pigeonpea hybrid breeding." *Frontiers in Plant Science* 6:793.

Saxena, Swati, Sarika Sahu, Tanvi Kaila, Deepti Nigam, Pavan K. Chaduvla, A. R. Rao, Sandhya Sanand, N. K. Singh, and Kishor Gaikwad. 2020. "Transcriptome profiling of differentially expressed genes in cytoplasmic male-sterile line and its fertility restorer line in pigeon pea (*Cajanus cajan* L.)." *BMC Plant Biology* 20:1-24.

Saxena, Swati, Tanvi Kaila, Pavan K. Chaduvula, Archana Singh, N. K. Singh, and Kishor Gaikwad. 2019. "Novel chloroplast microsatellite markers in pigeonpea (*Cajanus cajan*L. Millsp.) and their transferability to wild *Cajanus* species." *Australian Journal of Crop Science* 13:185.

Schnable, Patrick S., and Roger P. Wise. 1998. "The molecular basis of cytoplasmic male sterility and fertility restoration." *Trends in Plant Science* 3:175-180.

Seth, Purnima, Aniruddha P. Sane, Pravendra Nath, and Prafullachandra V. Sane. 1996. "Molecular characterization of mitochondrial genomes of rice lines containing wild abortive (WA) male sterile and fertile cytoplasms." *Journal of Plant Biochemistry and Biotechnology* 5:75-82.

Shinjo, Choyu. 1966. "Cytoplasmic-genetic male sterility in cultivated rice, Oryza sativa LI Fertilities of $F_1$, $F_2$ and offsprings obtained from their mutual reciprocal backcrosses and segregation of completely male sterile plants." *Japanese Journal of Breeding* 16:179-180.

Shull, George H. 1908. "The composition of a field of maize." *Journal of Heredity* 1:296-301.

Singh, B. D. 2005. *Plant Breeding: Principles and Methods*. New Delhi: Kalyani Publishers.

Singh, Hari P., and H. C. Lohithaswa. 2006. "Sorghum." In *Cereals and Millets*, 257-302. Springer, Berlin, Heidelberg.

Singh, Mahipal, and Gregory G. Brown. 1993. "Characterization of expression of a mitochondrial gene region associated with the Brassica

"Polima" CMS: developmental influences." *Current Genetics* 24:316-322.

Singh, Nagendra K., Deepak K. Gupta, Pawan K. Jayaswal, Ajay K. Mahato, Sutapa Dutta, Sangeeta Singh, Shefali Bhutani et al., 2012. "The first draft of the pigeonpea genome sequence." *Journal of Plant Biochemistry and Biotechnology* 21:98-112.

Singh, Vikas K., Rachit K. Saxena, and Rajeev K. Varshney. 2017. "Sequencing Pigeonpea Genome." In *The Pigeonpea Genome*, edited by Rajeev K. Varshney, Rachit K. Saxena and Scott A. Jackson, 93-97. Springer, Cham.

Sinha, Pallavi, K. B. Saxena, Rachit K. Saxena, Vikas K. Singh, V. Suryanarayana, C. V. Sameer Kumar, Mohan AVS Katta et al., 2015. "Association of *nad7a* gene with cytoplasmic male sterility in pigeonpea." *The Plant Genome* 8:1-12.

Song, Jiasheng, and Charles Hedgcoth. 1994. "A chimeric gene (*orf256*) is expressed as protein only in cytoplasmic male-sterile lines of wheat." *Plant Molecular Biology* 26:535-539.

Sotchenko, V. S., A. G. Gorbacheva, and N. I. Kosogorova. 2007. "C-type cytoplasmic male sterility in corn." *Russian Agricultural Sciences* 33:83-86.

Spassova, Mariana, Françoise Moneger, Christopher J. Leaver, Peter Petrov, Atanas Atanassov, H. John J. Nijkamp, and Jacques Hille. 1994. "Characterisation and expression of the mitochondrial genome of a new type of cytoplasmic male-sterile sunflower." *Plant Molecular Biology* 26:1819-1831.

Srikanth, Sandhya, Rachit K. Saxena, M. V. Rao, Rajeev K. Varshney, and Nalini Mallikarjuna. 2015. "Development of a new CMS system in pigeonpea utilizing crosses with *Cajanus lanceolatus* (WV Fitgz) van der Maesen." *Euphytica* 204:289-302.

Sultana, R., M. I. Vales, K. B. Saxena, A. Rathore, S. Rao, S. K. Rao, M. G. Mula, and R. V. Kumar. 2013. "Waterlogging tolerance in pigeonpea (*Cajanus cajan* (L.) Millsp.): genotypic variability and identification of tolerant genotypes." *Journal of Agricultural science* 151:659-671.

Sundaram, R. M., B. Naveenkumar, S. K. Biradar, S. M. Balachandran, B. Mishra, M. IlyasAhmed, B. C. Viraktamath, M. S. Ramesha, and N. P. Sarma. 2008. "Identification of informative SSR markers capable of distinguishing hybrid rice parental lines and their utilization in seed purity assessment." *Euphytica* 163:215-224.

Tang, H. V., W. Chen, and D. R. Pring. 1999. "Mitochondrial *orf107* transcription, editing, and nucleolytic cleavage conferred by the gene *Rf3* are expressed in sorghum pollen." *Sexual Plant Reproduction* 12:53-59.

Tester, Mark, and Peter Langridge. 2010. "Breeding technologies to increase crop production in a changing world." *Science* 327:818-822.

Tikka, S. B. S., L. D. Parmar, and R. M. Chauhan. 1997. "First record of cytoplasmic-genic male-sterility system in pigeonpea and its related wild species." *Journal of Plant Physiology* 137:64-71.

Tingdong, Fu, Yang Guangsheng, and Yang Xiaoniu. 1990. "Studies on "three line" polima cytoplasmic male sterility developed in *Brassica napus* L." *Plant Breeding* 104:115-120.

Tuteja Reetu, Rachit K. Saxena, Jaime Davila, Trushar Shah, Wenbin Chen, Yong-Li Xiao, Guangyi Fan et al., 2013. "Cytoplasmic Male Sterility-Associated Chimeric Open Reading Frames Identified by Mitochondrial Genome Sequencing of Four *Cajanus* Genotypes". *DNA Research* 20:485-495.

Ui, Hajime, Mohammad Sameri, Mohammad Pourkheirandish, Men-Chi Chang, Hiroaki Shimada, Nils Stein, Takao Komatsuda, and Hirokazu Handa. 2015. "High-resolution genetic mapping and physical map construction for the fertility restorer *Rfm1* locus in barley." *Theoretical and Applied Genetics* 128:283-290.

Vales, M. I., GV Ranga Rao, H. Sudini, S. B. Patil, and L. L. Murdock. 2014. "Effective and economic storage of pigeonpea seed in triple layer plastic bags." *Journal of Stored Products Research* 58:29-38.

Varshney, R. K., R. V. Penmetsa, S. Dutta, P. L. Kulwal, R. K. Saxena, S. Datta, T. R. Sharma et al. 2010. "Pigeonpea genomics initiative (PGI): an international effort to improve crop productivity of pigeonpea (*Cajanus cajan* L.)." *Molecular Breeding* 26:393-408.

Varshney, Rajeev K., Wenbin Chen, Yupeng Li, Arvind K. Bharti, Rachit K. Saxena, Jessica A. Schlueter, Mark TA Donoghue et al., 2012. "Draft genome sequence of pigeonpea (*Cajanus cajan*), an orphan legume crop of resource-poor farmers." *Nature Biotechnology* 30:83-89.

Vedel, Fernand, M. Pla, V. Vitart, S. Gutierres, and Ph Chetrit. 1994. "Molecular basis of nuclear and cytoplasmic male sterility in higher plants." *Plant Physiology and Biochemistry (Paris)* 32:601-608.

Venkateswarlu, S., A. R. Reddy, R. Nandan, O. N. Singh, and R. M. Singh. 1981. "Male-sterility associated with obcordate leaf shape in pigeonpea." *International Pigeonpea Newsletter* 1:16.

Verma, M. M., and P. S. Sandhu. 1995. "Pigeonpea hybrids: historical development, present status and future prospective in Indian context. In *Hybrid Research and Development*, edited by Mangala Rai and S. Mauria, 121-137, Indian Society of Seed Technology, IARI, New Delhi, India.

Vinod, K. K. 2005. "Cytoplasmic genetic male sterility in plants. A molecular perspective." *Proceedings of the training programme on Advances and Accomplishments in heterosis breeding.* Tamil Nadu Agricultural University, Coimbotore, India.

Wan, Zhengjie, Bing Jing, Jinxing Tu, Caozhi Ma, Jinxiong Shen, Bin Yi, Jing Wen et al., 2008. "Genetic characterization of a new cytoplasmic male sterility system (hau) in *Brassica juncea* and its transfer to *B. napus*." *Theoretical and Applied Genetics* 116: 355-362.

Wang, Kun, Feng Gao, Yanxiao Ji, Ying Liu, Zhiwu Dan, Pingfang Yang, Yingguo Zhu, and Shaoqing Li. 2013. "ORFH 79 impairs mitochondrial function via interaction with a subunit of electron transport chain complex III in Honglian cytoplasmic male sterile rice." *New Phytologist* 198:408-418.

Wanjari, K. B., A. N. Patil, M. C. Patel, and J. G. Manjaya. 2000. "Male sterility derived from *Cajanus sericeus*× *Cajanus cajan*: Confusion of cytoplasmic male sterility with dominant genic male sterility." *Euphytica* 115:59-64.

Wanjari, K. B., A. N. Patil, Prema Manapure, J. G. Manjayya, and Manish Patel. 1999. "Cytoplasmic male sterility in pigeonpea with cytoplasm from *Cajanus volubilis*." *Annals of Plant Physiology* 13:170-174.

Wettstein, F. V. 1924. "Morphologie und Physiologie des Formwechsels der Moose auf genetischer Grundlage. I." *Zeitschrift für induktive Abstammungs-und Vererbungslehre* 33:1-236.

Whitford, Ryan, Delphine Fleury, Jochen C. Reif, Melissa Garcia, Takashi Okada, Viktor Korzun, and Peter Langridge. 2013. "Hybrid breeding in wheat: technologies to improve hybrid wheat seed production." *Journal of Experimental Botany* 64:5411-5428.

Yamamoto, Masayuki P., Hiroshi Shinada, Yasuyuki Onodera, Chihiro Komaki, Tetsuo Mikami, and Tomohiko Kubo. 2008. "A male sterility-associated mitochondrial protein in wild beets causes pollen disruption in transgenic plants." *The Plant Journal* 54:1027-1036.

Yamamoto, Masayuki P., Tomohiko Kubo, and Tetsuo Mikami. 2005. "The 5′-leader sequence of sugar beet mitochondrial *atp6* encodes a novel polypeptide that is characteristic of Owen cytoplasmic male sterility." *Molecular Genetics and Genomics* 273:342-349.

Yang, Jinghua, Xunyan Liu, Xiaodong Yang, and Mingfang Zhang. 2010. "Mitochondrially-targeted expression of a cytoplasmic male sterility-associated *orf220* gene causes male sterility in *Brassica juncea*." *BMC Plant Biology* 10:231.

Yang, Soojung, Toru Terachi, and Hiroshi Yamagishi. 2008. "Inhibition of chalcone synthase expression in anthers *of Raphanus sativus* with Ogura male sterile cytoplasm." *Annals of Botany* 102:483-489.

Zabala, Gracia, Susan Gabay-Laughnan, and John R. Laughnan. 1997. "The nuclear gene Rf3 affects the expression of the mitochondrial chimeric sequence R implicated in S-type male sterility in maize." *Genetics* 147:847-860.

Zhang, Huamin, Junqing Wu, Zihui Dai, Meiling Qin, Lingyu Hao, Yanjing Ren, Qingfei Li, and Lugang Zhang. 2017. "Allelism analysis of *BrRfp* locus in different restorer lines and map-based cloning of a fertility restorer gene, *BrRfp1*, for *pol* CMS in Chinese cabbage (*Brassica rapa* L.)." *Theoretical and Applied Genetics* 130:539-547.

Zhang, Ming-Fang, Li-Ping Chen, Bing-Liang Wang, Jing-Hua Yang, Zhu-Jun Chen, and Yutaka Hirata. 2003. "Characterization of *atpA* and *orf220* genes distinctively present in a cytoplasmic male-sterile line of tuber mustard." *The Journal of Horticultural Science and Biotechnology* 78:837-841.

Zhao, T. J., and J. Y. Gai. 2006. "Discovery of new male-sterile cytoplasm sources and development of a new cytoplasmic-nuclear male-sterile line NJCMS3A in soybean." *Euphytica* 152:387.

In: *Cajanus cajan*
Editor: Donald S. Wilkes
ISBN: 978-1-53619-134-9


*Chapter 2*

# PHARMACOLOGICAL AND MEDICINAL PROPERTIES OF *CAJANUS CAJAN*

***Suman Pahal[1,*], Shruti Sinha[2,†], Manish Jangra[3,‡], Sonia Goel[4] and Sapna Grewal[1]***

[1]Department of Bio and Nano Technology, Guru Jambheshwar University of Science & Technology, Hisar, Haryana, India
[2]ICAR-NIPB, Indian Agricultural Research Institute, PUSA, New Delhi, India
[3]Department of Botany and Plant Physiology, Chaudhary Charan Singh Haryana Agricultural University, Hisar, Haryana, India
[4]Faculty of Agricultural Sciences, SGT University, Gurugram, Haryana, India

## ABSTRACT

Being rich in protein, dietary fibers and other nutrient contents, pulses are rightly called as poor people's meat and they make a well-

* Corresponding Author's E-mail: suman.pahalsg04@gmail.com.
† Corresponding Author's E-mail: shrutiagri22@gmail.com.
‡ Corresponding Author's E-mail: manish.jangra.330@gmail.com.

balanced diet in combination with cereals. Pulses have several health benefits besides being nutritionally rich. These are contributed by large number and diversity of phytochemicals present as secondary metabolites like phenolics, phytates, lectins, flavonoids, saponins, phytosterols etc. which are reported to have pharmacological properties. Among legumes, *Cajanus cajan* (also commonly known as pigeon pea) is one of the most valuable, versatile, nutritional and health promoting grain crop of the world. Although primarily grown as an agricultural crop for food, it is the only woody legume in the world with affirmed medicinal properties including anti-inflammatory, anti-diabetic, anti-cancerous, antioxidant, hepato-protective, anti-helminthic etc. Its leaf extract can be used for treatment of skin infections, measles, jaundice and is also an antidote against constipation, food poisoning and stomatitis. In recent years, considerable progress and advances have been achieved in studying potential medicinal and biological applications of pigeon pea and establishing it as a medicinal drug. Different forms of aqueous, alcoholic as well as other organic solvent derived extracts from various tissues primarily leaves and root have been isolated to harness these phytochemicals. These have been known to act either directly or in combination with either other phytochemicals or trade drugs in alleviating certain disorders, diseases and are thus promising candidates for drug discovery. This chapter includes a detailed focused account of the presence of various organic phytocompounds in it such as stilbenes, flavonoids, polyphenols, cajaninstilbene acid, luteolin, apigenin, isorhamnetin etc. and the medicinal/pharmacological properties associated. With added advantage owing to its edibility in combination with excellent medicinal properties, *C. cajan* have the potential to be used as a pharmaceutical supplement in near future.

**Keywords**: *Cajanus cajan*, medicinal plant, anticancer, antidiabetic, phytochemicals

## 1. INTRODUCTION

Pulses constitute an important food source and are well known for their high nutrient composition. In India, pigeon pea (*Cajanus cajan)* is a popular kharif legume pulse, which is grown on a wide scale under rainfed conditions. In India, we have 3.9 million hectares of land under pigeon pea cultivation (72% of the total area worldwide; FAO, 2018) contributing to almost 85% of world's total crop production (Jeevarathinam and Chelladurai 2020; Singh and Vaishampayan 2017; Sarkar et al. 2018). Pigeon pea stands out from other leguminous crops because of its specific unique traits such as stress tolerance, variable maturity time, natural out-crossing characteristics etc. (Saxena et al. 2019). It is nutritionally loaded with high quality proteins that are particularly rich in methionine, lysine and tryptophan. It holds a special significance as far as India is concerned because the poorer section of our society largely depends on such cheap sources of protein to meet their nutritional needs (Tiwari & Shivhare 2016). In developing economies worldwide, an ideal combination of cereals with legumes makes a nutritious, balanced and health enhancing meal as pulses are affluent with vitamins [riboflavin (B2), thiamine (B1), niacin (B3), folate], minerals (calcium, potassium, phosphorus, iron and magnesium) and many phyto-compounds. Pulses are hence rightly considered as 'poor people's meat' as they are cheap yet a significant source of protein (Basker et al. 2016; Venkidasamy et al. 2019).

Herbal or medicinal folk plants are gaining a lot of attention these days because of presence of multiple bioactive components and their role as pharmaceutical remedies to cure diseases. *Cajanus cajan*, a representative genera of legume family, is also an herbal plant owing to its innumerable and significant therapeutic traditional uses. Numerous literature studies have supported the fact that different plant parts of pigeon pea can help over with avoiding various common diseases (Grover et al. 2001; Singh & Basu 2012; Talari et al. 2018). Although these qualities are present in all plant parts, seeds are of immense importance being protein and fiber rich, having low glycemic index, high number of antioxidants, good water binding capacity and prominent fat absorption property (Keshav 2015; Talari et al. 2018).

Literature summarizing information on major phytocompounds present in this crop which provide medicinal attributes to it such as antioxidant, antidiabetic, antimicrobial is available (Figure 1) (Kong et al. 2009; Ezike et al. 2010; Boye et al. 2010; Tosh & Yada 2010; Pratima et al. 2011; Campos-Vega et al. 2013). Traditionally, *Cajanus cajan* leaves were used for treating ulcers, meal prompted hypercholesterolemia & malaria whereas, its seeds were used to treat hepatitis, diabetes, plasmodial diseases and cough (Mathew et al. 2017). On the other hand, its roots, kernels and leaves were used to cure jaundice, measles, diabetes, ischemic necrosis of femoral head, osteoporosis, fungal infections etc. (Huang et al. 2016). Pigeon pea also has the potential to prevent and cure skin irritations, sores, hepatitis, bladder stones, dysentery, cough etc. (Grover et al. 2002; Grover et al. 2001; Duker-Eshun et al. 2004; Kong et al. 2011; Gu et al. 2018). In Rajasthan, (India) people take its leaf juice orally to get relief from parasites, food poisoning and constipation and use the seed paste as an energy stimulant, treatment of gingivitis as well.

In Chinese tradition, *Cajanus cajan* leaves were reported to be used in relieving pain and treatment of wounds, malaria, hypercholesterolemia and even bedsores. Flavonoids and stilbenes are the two major phytocompounds present imparting various herbal therapeutic properties in the plants of genus *Cajanus* (Wu et al. 2009). Although considerable success has been achieved in unfolding phytochemistry of *Cajanus cajan* regarding its biological and pharmacological applications but in order to exploit the maximum potential of phytocompounds, a complete understanding of chemical, physical and biological properties of bioactive components and their health-enhancing effect has to be achieved (Pal et al. 2011; Syed and Wu 2018).

A number of conventional over the counter medications are available that are being used to treat various sicknesses but these are associated with several side effects affecting health in a negative way and hence the demand of plant-based drugs arises (Mathew et al. 2017). Due to presence of medicinally important bioactive constituents, cost effectiveness of plant-based drugs, enhanced therapeutic effects, minimal side effects on health and potency to address multiple issues, herbal sources are gaining importance and attention of researchers for avoiding and curing a number of health ailments.

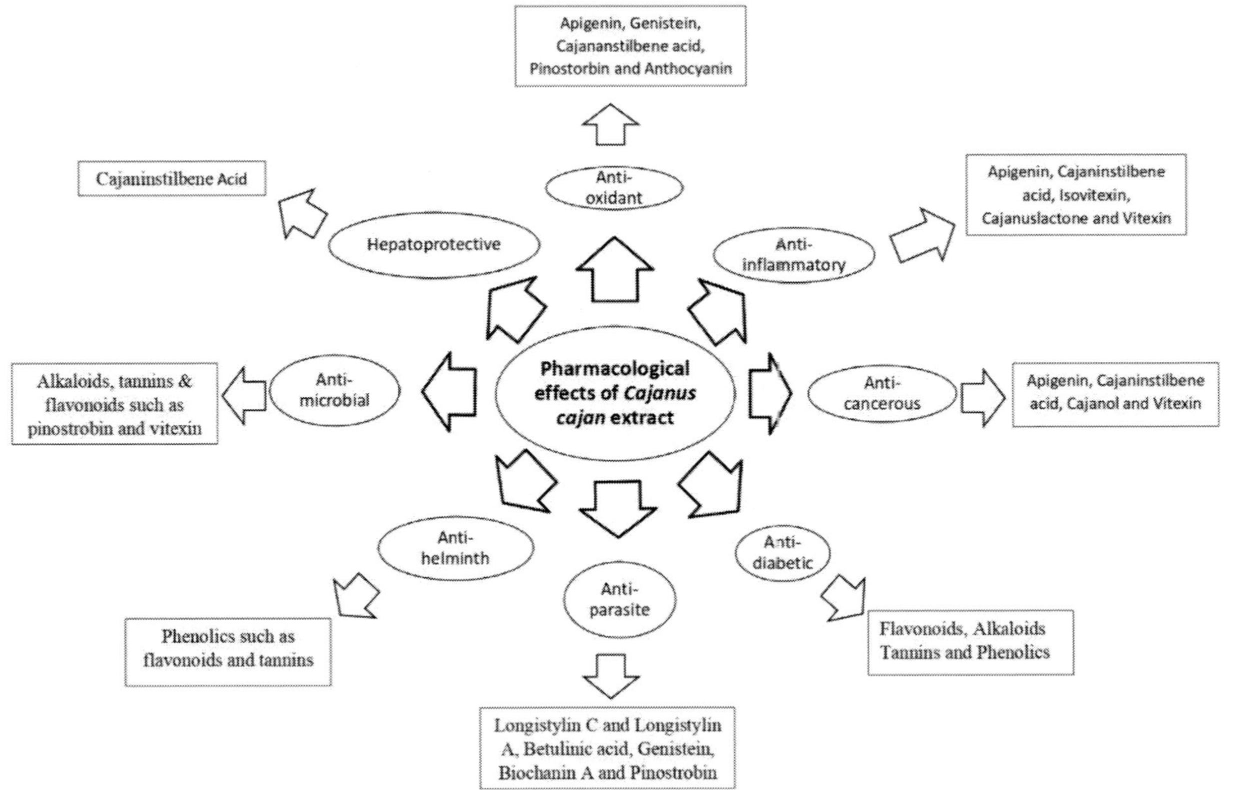


Figure 1. Major phytochemicals found in *Cajanus cajan* and their associated medicinal property.

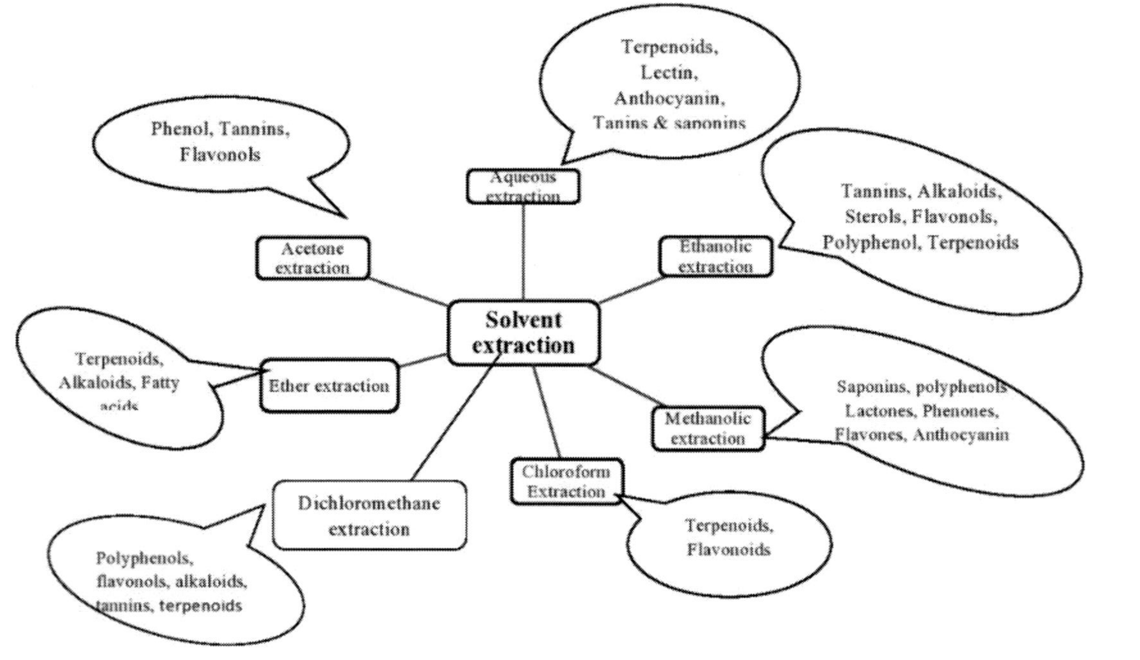


Figure 2. Major extraction methods used for isolation of phytochemicals from *C. cajan.*

## 2. Active Components and Biochemical Compounds Found in *Cajanus cajan*

Phyto-compounds are biologically active, plant based chemical compounds/secondary metabolites known to have protective, medicinal and curative properties. They have the potential to treat various human illnesses since time immemorial by boosting immune system. A number of extraction methods using different extraction media like ethanol, methanol, water, ethyl acetate, n-hexane, petroleum ether etc. have been designed to isolate these bioactive compounds from plants (Figure 2). Leaves contain ample amount of these active compounds and therefore the leaves of *Cajanus cajan* are arranged as infusion for treatment of hepatitis, anaemia, urinary infections, diabetes, genital, other skin irritations and yellow fever while floral decoctions got potential cure for coughs, dysentery, bronchitis, menstrual disorders and pneumonia.

Cajaninstilbene acid (CSA), one of the vital attention seeker agile constituent, isolated from pigeon pea leaves, exhibits strong antioxidation activities as comparable to resveratrol, which is referred as a regular antioxidant by medical practitioners (Iacopini et al. 2008). CSA is also hypoglycemic by nature (Hopp and Inman 2002), antibacterial (Kong et al. 2010), has antitumor activity and deoxyribose nucleic acid (DNA) damage protection role too (Huang et al. 2016; Fu et al. 2015). *Cajanus cajan* aqueous extracts have been recorded for treatment of various health illnesses like diabetes, dysentery, hepatitis and measles because of the availability of cajaminose & phenylalanine (Iwu et al. 1988; Onah et al. 2002). *Cajanus cajan* root and leaf extracts on bioactivity-guided fractionation were found to contain 8 main bioactive compounds: biochanin A, betulinic acid, genistein, 2′-hydroxygenistein, cajanol, longistylin A, longistylin C and pinostrobin (Duker-Eshun 2004).

Isovitexin and vitexin (flavones) are reported to have therapeutic applications in many studies (Nix, Paull and Colgrave 2015). Vitexin is anti-inflammatory, antiplasmodic, anti-cancerous, hypotensive, cardioprotective, anti-diabetic and memory enhancer in nature (Agnese et al. 2001; Prabhakar et al. 1981; Picerno et al. 2003; Bramati et al. 2003). Desaspidinol is found to be most effective for arthritis, diabetes and breast cancer, validating it as the most promising phytocompound that can be used as a Phyto drug. Likewise, Quinazolin-2-ol was established having capabilities against diabetes, arthritis, measles and malaria & its derivatives are known for their therapeutically important properties against tumor, bacteria, inflammation, cancer, hypertension, hyperlipidemic, malaria etc. Stilbenes & flavonoids both seem to be present in higher amount in *C. cajan* leaves in comparison to the pods. The presence of 2′-hydroxy genistein, 2′-2′-methyl cajanone, cajanin, isoflavones etc., in *C. cajan* imparts antioxidant characteristics (Pal et al. 2011). Almost 320 phyto-compounds were found using GC-MS which were narrowed down to 242 after removing duplicates from various plant extracts and put through cytotoxicity projection of drug with help of ADMET analysis that identified significant number of promising potential drug candidate towards many human diseases like breast cancer, rheumatoid arthritis, diabetes, malaria, sickle cell disease and measles (Mathew et al. 2017).

Phenolic compounds of this legume crop (Flavonoids, Tannins, Alkaloids, Saponins, Cyanogenic glycoside, Anthocyanin) are found to contain coumarin cajanuslactone as bioactive constituent which is responsible for treating sores, diabetes, bedsores, skin, jaundice, measles (Pal et al. 2011). Effects of various phytocompounds on health and potential treatment strategies can be established by understanding interaction of these phytocompounds in human body. *Cajanus cajan* extracts contain ample amount of beneficial phytocompounds or active constituents namely flavonoids like flavones, isoflavones, flavonones, chalcone and stilbenes (Ahmed et al. 2019) which are reported to impart therapeutic properties and bioactivities (Kong et al. 2010; Wu et al. 2009). Lower-energy collision-induced dissociation tandem mass spectrometry (CID-MS/MS) analysis lead to identification of ten analytes including two stilbenes, two isoflavonoids and six flavonoids. In pigeon pea, LC-MS/MS lead to separation and identification of these three categories of proactive constituents i.e., isoflavonoids, stilbenes and flavonoids for the first time (Liu et al. 2010). Stilbenes are one of main active phytocompounds in *C.*

*cajan* which caused effects on lipid metabolism as confirmed by investigating their role in mice grown on a high-fat diet (Luo et al. 2008). Although all of the plant parts contain therapeutically active constituents of medicinal importance but leaves are still the most sought-after source of bioactive compounds including anthocyanins (Ade-Omowaye et al. 2015).

# 3. Pharmacological and Medicinal Implications of *Cajanus* Extract on Human Health

Potential health beneficial, medicinal properties affirmed with *Cajanus cajan* includes anti-microbial, anti-bacterial, anti-inflammatory, anti-cancer, hypocholesterolemia effects, anti-diabetic, antioxidant, neuroactive properties, anthelmintic, hepatoprotective etc. Since time immortal, pigeon pea excerpts such as quercetin, polyphenols, apigenin, luteolin, flavonoids, cajaninstilbene acid, isorhamnetin, etc. are used for treating diarrhea, jaundice, cough, sores, bronchitis, diabetes and bladder-stones with minimal side effects.

## 3.1. Antioxidant Activity of *Cajanus cajan*

Cell metabolism produce free oxygen or nitrogen associated radicals as byproduct, low to moderate concentration of these reactive species plays important roles in physiological processes from signal transduction to defense mechanism. But as their concentration increases, these becomes detrimental to cells causing oxidative stress resulting in degradation of cell components, protein, DNA, lipids, carbohydrates resulting in onset of cancer, rheumatoid arthritis, diabetes, cirrhosis etc. due to deficiency of enzymatic or non-enzymatic antioxidant defense system (Nahar et al. 2014; Rahman et al. 2016). Oxidative stress is one of main reason of diabetes and associated complications (Halliwell et al. 1989). Anti-oxidants (also referred to as free radical scavengers) are one of the most talked about compounds these days as they are known to prevent the damage caused by free-radicals to the cells. They work by acting as singlet oxygen quencher, metal chelating compounds or simply an electron/hydron donor. Plants based antioxidants are simple phytonutrients that are taken through plant-based diets in natural forms. Oxidative stress is linked directly to heart diseases, cancer and many inflammatory disorders so having anti-oxidants *via* food is one of the easiest ways towards health (El-Mawla and Husam 2011). Legumes are affluent origin of bio-active compounds with antioxidant capabilities and are thus defined as functional food. A total of four compounds- cajaninstilbene acid (CSA), pinostrobin, orientin & vitexin extracted from ethanolic extracts of *C. cajan* leaves have been observed to have considerable pharmacological and antioxidant activities (Wu et al. 2009; Pal et al. 2011). Ethanolic extracts of pigeon pea leaves analyzed by LC-MS/MS identified another compound, apigenin as a potent antioxidant (Gao et al. 2012). Uchegbu and Ishiwu (2015) identified antioxidant activity in extract of germinated pigeon pea seedlings. When this germinated pigeon pea extract was consumed by alloxan-induced diabetic rats, it caused lowering of blood glucose level (Uchegbu & Ishiwu 2015; Talari et al. 2015). Potent anti-oxidant activities have also been found to be exhibited by pigeon pea seed husk (Rani et al. 2014). While aqueous extract of the pulse showed maximum anti-oxidant activity (Mahitha et al. 2015), methanolic extracts of *Cajanus cajan* also possessed appreciable antioxidant activity (Ganga Raju et al. 2018). Also, ethanol extracts from roots are reported to be rich in polyphenols (cajanol, genistein, daidzein) and genistein could possibly be the main component in root extracts, with effective anti-oxidative activity reducing ROS by enhancing SOD and CAT mainly by activating nuclear factor (NF) erythroid 2-related factor 2/antioxidant protein heme oxygenase-1 and inhibiting kappa B (NF-κB) signaling (Thuy-Lan Thi et al. 2010).

The role of proteins isolated from pigeon pea seeds as anti-oxidants was studied on $H_2O_2$-induced cellular damage (Muangman, Leelamanit & Klungsupaya 2011). ABTS, FRAP and DPPH assays were used for evaluation of antioxidant potential of *Cajanus cajan* correlating the phenolic contents and relative antioxidant activity (Xu & Chang 2011;

Marathe et al. 2011). DPPH based radical scavenging assay of leaf extracts and its isolated compounds (pinostrobin, cajaninstilbene acid, orientin, vitexin, ethyl acetate, petroleum ether, n-butanol and water fractions) observed a concentration-dependent antioxidant activity. Results concluded that pigeon pea leaf extracts are source of potential antioxidants and hence the crop can have good applications to medicine industry (Wu et al. 2009). 2-hydroxy genistein, 2'-2' methyl cajanone, cajanin, cahanones, isoflavones etc., are also reported to have antioxidant characteristics (Kong et al. 2010; Ganga Raju et al. 2018).

## 3.2. The Anti-Inflammatory Role

Inflammation is body response to signal the immune system against foreign invaders and is thus physiologically crucial to our survival. But chronic inflammation can be a problem and is linked to many common diseases such as rheumatoid arthritis, lupus and heart stroke. According to National Cancer Institute, chronic inflammation can cause DNA damage, leading to some forms of cancer too. Steroidal anti-inflammatory drugs (SAIDs) and non-steroidal anti-inflammatory drugs (NSAIDs) are commonly used to alleviate inflammation, but these are associated with several side effects affecting health adversely (Ganga Raju et al. 2018). NSAIDS block cyclo-oxygenase that prevents synthesis of prostaglandins which is responsible for inflammation. Though there is no magic diet to cure inflammation but the flare up frequency and severity can be controlled substantially using natural anti-inflammatory foods. There are evidences suggesting that bioactive constituents of *cajanu cajan* have a significant role in modifying the gut microbiota and thus help in reducing inflammation. In response to any mechanical injury, burn, allergens, microbial infections, protein denaturation or any other kind of noxious stimulus, body's immunological defense mechanism induces inflammation at that site (Menichini et al. 2009; Mueller et al. 2010; Leelaprakash and Mohan Dass 2010) and to treat these inflammatory disorders, various anti-inflammatory drugs or medicines are used in spite of their adverse side effects and thus, there comes desire for plant-based agents in place of chemical pharmaceutics having little or negligible adverse health effects (Sosa et al. 2002).

Quite a number of studies done showing anti-inflammatory effects of *Cajanus cajan* leaf extracts having Cajaninstilbene acid (CSA). CSA and its associated derivatives have affirmed potential antioxidant and anti-inflammatory activities confirmed by studying their effects in lipopolysaccharides stimulated mice macrophages where CSA strongly blocked the release of nitric oxide and inflammation causing factors TNF-α & IL-6, thus reducing inflammation (Huang et al. 2016; Syed and Wu 2018). In a similar study, secretion of inflammatory cytokines, TNF-α, IL-1β, and IL-6 has been found to be suppressed by almost 50% in hydrogen peroxide treated macrophages using the ethanol extracts of cyanidin-3-monoglucoside and pigeon pea extracted anthocyanin (Lai et al. 2012). Carrageenan-induced rat paw oedema presents a simple, routine, experimental animal model to evaluate anti-inflammatory activities at the site of pain and inflammation (Sugishitha et al. 1981). The hexane-based extract of pigeon pea seeds suppressed TNF-α & IL-6 in the serum of rat with inflammation induced by carrageenan (Hassan et al. 2016; Syed and Wu 2018).

The lupeol present in the extract exert its activity by targeted inflammatory signaling pathways. Poly phenol contents, flavonoids, saponins are considered responsible for anti-inflammatory nature of methanolic extracts of *Cajanus cajan* (MECC) as they can bind to cations, thereby inhibiting protein membrane denaturation which is also one reason of inflammation (Leelaprakash & Mohan Dass 2010). In lipopolysaccharides stimulated J774A.1 cells & RAW264.7 cells, cajanolactone extracted from pigeon pea leaves have been found to lower the secretion of inflammatory factor TNF- and IL-1β to a significant level and thereby reducing inflammation. These studies showed that *Cajanus cajan* extracted natural bioactive compounds and their associated derivatives could be some excellent agents to be used as anti-inflammatory molecules. It has been confirmed by both *in vivo* and *in vitro* studies done on zebra fish and macrophage cell models respectively that CSA and associated derivatives are potential anti-inflammatory molecules. Therefore, these molecules can be used as functional food, complementary medicine, or strong candidates to develop novel drugs against inflammation (Huang et al. 2016).

It has been concluded that for the generation of bioactive phenolic compounds, especially the health-promoting compound cajaninstilbene acid, PPHRCs (Pigeon pea hairy root cultures) could be an important substitute to pigeon pea natural resources (Jiao et al. 2020). Peroxisome proliferator-activated receptor gamma (PPARγ) helps in treating cancer, inflammation & inflammation-related disorders to a remarkable extent exhibiting its therapeutic properties as confirmed by studying anti-inflammatory and cytotoxic activities of *C. cajan* and its bioactive constituents on three cancerous lines namely, human cervical adeno carcinoma cell line (HeLa), the human colorectal adeno carcinoma cell line (CaCo-2) and the human breast adeno carcinoma cell line (MCF-7). High anti-inflammatory and cytotoxic response of *C. cajan* based phytochemicals was noted on the lipopolysaccharide-stimulated macrophages and cancer lines proving PPARγ activity *in vitro*. Although 3 phytocompounds namely orientin, pinostrobin and vitexin were concluded to be mainly responsible behind this activity, novel PPARγ activators identified were Cajaninstilbene acid and pinosylvin monomethyl ether. All these studies suggest that pigeon pea leaves might be an herbal remedy towards effectively treating & inhibiting inflammation or other such ailments (Schuster et al. 2016).

## 3.3. Role as Tumour Suppressor Agent

Plant-derived secondary metabolites are well-known to have acted against different types of cancer (Baskar, Park, & Nile 2016). Isoflavonoids, flavonoids, flavones are some of the secondary metabolites in pigeon pea which plays main role in imparting anticancerous properties to the plant. Genistein is one of the main isoflavone which has been proved to have a role in breast cancer chemoprevention. It shows its effect by inducing apoptosis through mitochondrial-dependent pathways at high concentrations (Luo M et al. 2010). Constantinou and his co-workers (1998) observed that 150 M genistein treatment for 48 hours lead to 57.5% reduction in apoptotic MCF-7 cells. There are several phytocompounds that act against tumor including genistin, but cajanol is more into focus these days as far as cancer therapy is concerned. Cajanol [5-hydroxy-3-(4-hydroxy-2-methoxyphenyl)-7-methoxychroman-4-one] is an effective cytotoxic phyoalexin against tumor cells extracted from pigeon pea roots. It results in apoptosis in human breast cancerous cells and therefore it could be a potential candidate for anticancerous drug designing (Anwar et al. 2010; Liu et al. 2010; Zhao et al. 2013). In addition of being anti-cancerous, Cajanol has antifungal, antimicrobial and antiplasmodial activities too. Although both Cajanol and genistin have similar chemical structure but cajanol is observed to cause higher percentages (approx. 80%) of apoptosis in MCF-7 cells at comparatively lower concentrations (64 M) (Luo et al. 2010).

In pigeon pea, alkaloids too play a role in inducing response against stress and causing apoptosis in human breast cancer cell. The methanol excerpt of the plant has been found to have shown cytotoxicity towards three cancer cell lines, namely human lung carcinoma cell line COR-L23, human amelanotic melanoma (C32) and human breast adenocarcinoma cell line (MCF-7) (Ashidi et al. 2010; Talari et al. 2015). A number of compounds like genistin, genistein, α amyrin, hexadecanoic acid, pinostrobin, β-sitosterol, longistylin C and longistylin A are found to be responsible for anticancerous activity of *Cajanus cajan* root extracts (Ganga Raju et al. 2018). Its cytotoxic activities towards MCF-7 breast cancer cells have been evaluated in detail. Several assessments like DNA fragmentation and nuclear changes assay, ROS generation, expression of enzymes involved in apoptosis like caspase-9, caspase-3, cytochrome c, bax, bcl-2 were done to understand the mechanism of cell growth inhibition of cajanol in MCF-7 cells. Cajanol was seen to arrest cell cycle in the G2/M phase, thereby inducing apoptosis through ROS-mediated mitochondria-dependent pathway demonstrating the cytotoxic activity of cajanol towards cancer cells *in vitro*. This established the DNA synthesis inhibiting activity of cajanol at the G2/M boundary. The inhibition of growth of MCF-7 cells was in a time and dose-dependent manner.

Both plants and fungi are mutually beneficial to each other. Endophytic fungus is like pool of natural, valuable secondary metabolites and bioactive constituents which have found potential applications in several fields like pharmaceutical, agrochemical and food industries (Tan and Zou 2001; Kaul et al. 2012; Chen et al. 2016). Many such fungus has been well documented and demonstrated to be associated with several crops including pigeon pea which could be used as an alternative source for exploiting promising bioactive compounds that can be used in plant-based drug designing like luteolin, apigenin and cajaninstilbene acid (Gao et al. 2012; Zhao et al. 2013, 2014). In this regard, a novel, important endophytic fungi named Hypocrea lixii, has been isolated from pigeon pea producing fungal anticancer agent- cajanol with stronger cytotoxic against human lung carcinoma cells (A549) *in vitro*. Taking this study as the base, we can find out some potential substitute methods for large-scale production of cajanol so as to meet the huge demand of anticancerous drugs. There are several marketed chemotherapeutical drugs available for breast cancer but these are not completely reliable due to increased resistance and severe side effects associated with them. This raise the need of other effective, reliable, novel alternatives for cancer treatment and thus new plant based or chemical products are urgent to discover, identify and isolate.

## 3.4. Role in Controlling and Curing Diabetes

Diabetes is a common but serious endocrine disorder impacting a very large population of about 100 million worldwide. In India, the number of people suffering from diabetes is increasing at a never seen rate and it is predicted that India is going to become diabetes capital of the world soon (Jacob and Narendhirakannan 2018). Although there are several medications or drugs available in the market showing hypoglycemic effects while alleviating diabetic symptoms, but these over-the-counter drugs are not cheap, have their own undesirable effects and sometimes known to cause complications like nephrological disorders, diarrhea, fatigue too. Oxidative stress which results in decreased antioxidant activity contributes largely in development of diabetes (Maritim et al. 2003; Palanisamy et al. 2011).

Nutritionally rich legumes with effective antioxidant potentials also possess anti-hyperglycemic activity and hence can play a role in the treatment of diabetes (Ashok et al. 2013; Dolui and Sengupta 2012). Sprouting of pulses including pigeon pea makes them much healthier and nutritious as germination enhances both quality and quantity of most secondary metabolites. These sprouted seeds are rich sources of bioactive compounds or constituents like alkaloids, flavonoids, tannins and phenolic which improvise the efficacy of pancreatic tissues by enhancing the insulin secretion or lowering the intestinal absorption of glucose (Kooti et al. 2016; Jacob and Narendhirakannan 2018). Thus, a meal containing sprouted pigeon pea is a good dietary supplement for controlling hyperglycemia & lipid peroxidation (El-Adawy et al. 2003). Germinated pigeon pea seed extracts when fed to alloxan-induced diabetic rats, reduced lipid peroxidation in liver tissue by 31.1% along with the reduction in blood glucose levels. Beta cells of pancreas are completely destroyed by alloxan in diabetic mice and *Cajanus cajan* leaf extract showed potential anti-diabetic activities probably because of insulin-like effect of extracts on tissues either by enhancing glucose metabolism or inhibiting gluconeogenesis (Nahar et al. 2014). Methanolic extracts of seed husks also illustrated free radical scavenging and anti-oxidant activities *in vitro* on hyperglycemic rats and mitigated starch induced postprandial glycemic excursions (Ashok et al. 2013).

It is thus also considered as an efficient and effective pharmacological plant which is commonly studied for treating diabetes and its associated malfunctions. MCCR (methanolic extracts of *Cajanus cajan* roots) has been proven to play effective role in managing diabetes because of its hypoglycemic nature and antioxidant potential (Nahar et al. 2014; Rahman et al. 2016). Crude methanol extract of *Cajanus cajan* seed husks worked similar to drug acarbose by eliminating starch-induced glycemic excursions and reducing glycemic load in mice (Tiwari et al. 2013). The hypoglycemic effect of crackers formed from sprouted pulse caused hypoglycemic effect in diabetic mice and reduced biochemical indices (Uchegbu and Ishwin 2016; Talari et al. 2015). Improvement in lipid profile and hypoglycemic effect resulting in decrease cardio metabolic risk has been observed in overweight type 2 diabetic patients when red meat diet was replaced with legumes proving legumes as a health beneficial herbal crop (Niazi et al. 2015).

### 3.5. Activity against Worms and Parasites

*Plasmodium* genus is responsible for diseases like malaria, parasitic infection etc. Pigeon pea has found application in treating malarial disease also. Root and leaf extract of *Cajanus cajan* have been identified for the presence of longistylin C and longistylin A, betulinic acid, genistein, biochanin A, cajanol 2′-hydroxygenistein and pinostrobin as important compounds with anti-malarial activity. The two stilbenes (longistylin C & longistylin A) and betulinic acid demonstrated a reasonably good *in vitro* activity towards the chloroquine-sensitive *P. falciparum* strain 3D7 (Duker-Eshun 2004; Orni et al. 2018). This study paves the way for possible application of *cajanus cajan* against malaria as we are still losing lives because of this disease every year.

A study to identify antihelminthic potential of *Cajanus* was conducted by Siddhartha et al. in 2009 employing earthworm (*Pheretima posthuma*) as the model system. Hydroalcoholic extracts were prepared from leaves and stems of *Cajanus* and were injected in earthworms to analyze the effect of these extracts in causing paralysis or death of the worm. The results were comparative to that of standard drug, piperazine citrate and the authors concluded that phenolics such as flavonoids and tannins are the responsible phytochemicals for this activity of the plant (Singh et al. 2009). Another study was conducted in 2015 to study the potential of organic-leaf extract of *Cajanus* in causing the mortality of encysted flukes. Flukeworm (*Fasciola hepatica*) causes Fasciolosis in cattle, a deadly hepatic disease.

The researchers tried different set of experiments to identify the lethal concentration, LC50 and postulated that the non-alkaloid nitrogen containing sterols present in *Cajanus* might have a role as anti-helminthic role. (Alvarez-Mercado et al. 2015).

## 3.6. Antimicrobial Activity against Pathogenic Bacterial and Fungal Strains

### *3.6.1. Antibacterial*

Supercritical fluid extraction (SFE)- mediated *cajanus* extracts showed *in-vitro* antimicrobial activity against *S. epidermidis, S. aureus* and *B. subtilis*. *Staphylococcus aureus* infection is a significant cause of soft-tissue, skin, bone, respiratory, endovascular disorders in humans. Zu et al. *in* 2010 characterized the SFE extract by HPLC-DAD and identified many flavonoids (pinostrobin, vitexin, orientin, isovitexin) and the stilbene-cajaninstilbene acid. Hydrogels isolated from pigeon pea seed husk were found to have antimicrobial activity against wound-harboring and/or aggravating microbes such as *Pseudomonas aeruginosa, Bacillus subtilis, Escherichia coli, Shigella dysentrae*. The hydrogel was way more effective than the standard medical formulations against *P. aeruginosa* which is responsible for delayed healing of wounds as it infects both acute as well as chronic wounds (Patil and Mastiholimath 2011).

Cajanuscoumarin- Cajanuslactone is reported to inhibit gram positive bacteria *S. aureus* growth (Kong et al. 2010). Seeds extract in organic solvents have been reported by Asma et al. (2015) to possess high antioxidant activities due to phenols and flavonoids and have potential to treat several chronic diseases even those that are bacterial causatives such as dysentery. In Nigeria, Bacillary dysentery due to *Shigella* and *Salmonella* species is extremely common causing high mortality. Methanolic extract of *Cajanus* leaves were shown to be effective in inhibiting both these microbes isolated from stools of infected children.

**Table 1. The list of major bioactive compounds found in *Cajanus cajan***

| Bioactive constituent | Plant part | Effect | References |
|---|---|---|---|
| Luteolin | Leaves | Antiviral | Cooksey et al. 1983, Jiao et al. 2020, Fu et al. 2008, Wei et al. 2013. |
| Apigenin | Leaves | anti-inflammatory, anticancer, antiviral, anti-oxidant | Wang and Huang 2004, Gao et al. 2012, Wei et al. 2013, Fu et al. 2008. |
| Genistein | Roots | Antioxidant, Insecticidal, anticancerous | Kang et al. 2003, Zhang et al. 2010, Jiao et al. 2020, Wu et al. 2008. |
| Genistin | Roots, etiolated stems | Antibacterial | Bhanumati et al. 1979b, Duker-Eshun et al. 2004, Jiao et al. 2020, Ganga Raju et al. 2018. |
| 2′-Hydroxygenistein | Roots, etiolated stems | Antioxidant | Duker-Eshun et al. 2004, Ingham 1976. |
| Cajanol | Roots | Anticancerous, antiplasmodial, antifungal | Luo et al. 2010, Wei et al. 2013, Liu et al. 2011. |
| Cajananstilbene acid | Leaves | DNA damage protective, Antioxidant, anti-inflammatory, anti-depressant, anti-tumor | Wu et al. 2011, Huang et al. 2016, Kong et al. 2009,2011, Fu et al. 2015. |
| Pinostorbin | Leaves | Antioxidant, anticancerous | Kong et al. 2009, Chen et al. 2014, Zhang et al. 2019, Wei et al. 2013. |
| Cajanuslactone | Leaves | Antibacterial, anti-inflammatory | Kong et al. 2010, Patel et al. 2014. |
| Cajaminose | Seeds | Sickle cell disease | Iwu et al. 1988, Mathew et al. 2017. |
| Desaspidinol | Seed | Breast cancer | Mathew et al. 2017. |
| Quinazolin 2-ol | Seed | Type-2 diabetes, measles, malaria | Mohamed, P.P.N. Rao 2017, Mathew et al. 2017. |
| Vitexin | Leaves, CCHRC | Anti-inflammatory, anti-tumor, anti-diabetic, hypotensive, cardioprotective | Prabhakar et al. 1981, Agnese et al. 2001, Jiao et al. 2020, Bramati et al. 2003, Picerno et al. 2003, Wu et al. 2009. |

**Table 1. (Continued)**

| Bioactive constituent | Plant part | Effect | References |
|---|---|---|---|
| Isovitexin | Root cultures, leaves | Anti-inflammatory, Antimicrobial | Jiao et al. 2020, Fu et al. 2007, Agnese et al. 2001. |
| Quercetin | Leaves, pod surface | Cytotoxic | Green et al. 2003, Zu et al. 2006, Wang et al. 2015, MAM Abo-Zeid et al. 2018. |
| Isorhamnetin | Leaves | Hypoglycemic | Zu et al. 2003, Sarkar et al. 2009, Jiao et al. 2020. |
| Cajanin | Seeds and etiolated stems | Antioxidant,hypocholesteric | Dahiya et al. 1984, Ingham 1976, Dahiya 1987, Mathew et al. 2017, luo et al. 2008. |
| Longistylin A | Leaves | Antiplasmodial, anticancerous, antimalarial, hypocholesterolemic | Duker Eshun et al. 2004, Luo et al. 2008. |
| Longistylin C | Leaves | Antiplasmodial, antidepressant, neuroprotective, anticancerous, Hypocholesterolemic | Liu et al. 2017, Ganga Raju et al. 2018, Luo et al. 2008. |
| Betulinic acid | Roots | Antiplasmodial | Duker Eshun et al. 2004. |
| Beta-sitosterol | Roots, leaves | Antitumor, antibacterial | Duker 2004. |
| Orientin | Leaves | Antioxidant, antiplasmodial, antibacterial | Wu et al. 2009, Jiao et al. 2020, Pal et al. 2011, Wei et al. 2013. |
| cajaChalcone | Leaves | Antimalarial | Ajaiyeoba et al. 2013. |
| Cajanone | Roots | Antifungal | Preston 1977, J Friend 2016, Dahiya 1991. |
| Biochanin A | Leaves, roots | Hypocholesterolemic | Jiao et al. 2020, Duker-Eshun et al. 2004, Wei et al. 2013a. |
| 2 hydroxygenistein | leaves | Antioxidant | Kong et al. 2017, Pal et al. 2011. |
| Anthocyanin | leaves | Antioxidant, anti-inflammatory | Lai et al. 2012. |
| phenylalanine | Seeds | Anti-sickling agent | Akojie et al. 1992, Ekeke and F.O. Shode 1990. |

| Bioactive constituent | Plant part | Effect | References |
|---|---|---|---|
| n-methyl-n-(4-(1-pyrrolidinyl)-2-butynyl)-2-aminoacetamide | Stem | Rheumatid arthritis | Mathew et al. 2017. |
| 3-hydroxy-4-methyoxymadelic acid | Seed | Rheumatoid arthritis | Mathew et al. 2017. |
| 2’,6’-dihydroxy-4’-methoxychalcone | Seed, leaf | Breast cancer | Mathew et al. 2017. |
| 2’,6’-dihydroxy-4’-methoxychalcone | Leaf, seed | Type-2 diabetes | Mathew et al. 2017. |
| 1,4,5-trimethylimidazole | Leaf | Malaria | Mathew et al. 2017. |
| 2,6-dimethoxyphenol | Stem | Sickle cell disease | Mathew et al. 2017. |
| 2-(4-nitrophenyl)-2-oxoethyl 3-phenyl propionate | Stem | Sickle cell disease | Mathew et al. 2017. |
| Hydroxybenzoic acid | Seeds | Antisickling agent | Akojie et al. 1992. |
| Cajanstilbene H | Leaves | promoted osteoblast differentiation in Human mesenchymal stem cells (hMSC), exhibiting a potential to alleviate osteoporosis, cytotoxic against cancerous lines. | Cai et al. 2015. |
| cajanstilbenoids A | Leaves | Cytotoxic | Zhang et al. 2017. |
| cajanstilbenoids B | Leaves | Cytotoxic | Zhang et al. 2017. |
| Cajanolactone A | Leaves | promoted osteoblast differentiation in Human mesenchymal stem cells (hMSC), anti-osteoporotic | Liu et al. 2019. |
| 2,6-dimethoxyphenol | Stem | Anti-sickling agents | Mathew et al. 2017. |
| 2-(4-Nitrophenyl)-2-oxoethyl 3-phenyl propionate | Stem | Anti-sickling agents | Mathew et al. 2017. |
| 2’-2’ methyl cajanone, | Leaves | Antioxidant | Pal et al. 2011. |
| Formanonetin (Biochanin B) | Leaves | Acts as Phytoalexin in etiolated stems | Wei et al. 2013. |

A patent was granted in China (2010) for the method of extracting monomer components such as stilbene acid, Semen Cajani lactone, pinostrobin, apigenin-8-C-glucoside, Saponaretin from crude *Cajanus cajan* leaf extract in organic solvents (ether, chloroform, acetic acid ethyl ester, n-butyl alcohol). Their study confirmed the inhibitory action of these phytochemicals against gram-positive bacteria which are sources of coliform diseases affecting large population in China. The pigeon pea leaf extract can be used for treating children pneumonia and upper respiratory tract infection caused by *staphylococcus aureus* and food poisoning caused by meat, milk, fish, eggs, food leftovers, bean jelly and glutinous rice cakes. Also, since the extract is non-toxic, cheap, easy to make and have possibly no chance of drug resistance to it (as is for penicillin and other antibiotics), it can be effectively used to make drugs in form of tablet, injection, capsule or even oral formulations. It clearly stated that the application of monomer ingredient in preparing medicament for resisting gram-positive bacteria that can be developed as the novel antibacterial medicine. [Patent Application Number: CN101485714A]. Likewise, Fagbohun et al. 2010 reported that the minerals such as Na, K, Ca, Mg, Zn and P might be the qualitative reason of antibacterial activities of *Cajanus* extract in both aqueous as well as methanolic medium against *Shigella sonnei* and *E. coli*.

Five human pathogenic bacteria isolated from Nigerian population viz., *Staphylococcus aureus, Escherichia coli, Salmonella typhi, Klebsiella pneumoniae and Proteus mirabilis* were inhibited by aqueous as well as organic solvent- mediated leaf extracts of Cajanus. Efficacy of organic solvent extracts were more inhibitory to *E. coli, S. aureus*, and *S. typhi*; while aqueous extract was inhibitory to *E. coli* and *S. aureus*. The other two bacterial isolates could not be inhibited and these *K. pneumoniae* and *P. mirabilis* found to have resistance to all the extracts (Okigbo and Omodamiro 2007). Leaf extract in organic solvent were used to study their antibacterial effect on Gram +ve and Gram –ve bacteria by Pratima et al. 2011. They found that ethanolic extract had a broad-spectrum antibacterial activity. The ethanolic extracts of *Cajanus cajan* L. showed highest sensitivity against *Klebsiella pneumoniae* and *Staphylococcus aureus*. The results support the use of *Cajanus cajan* L. in the cure of gastroenteritis, pneumonia and urinary tract infection. While alkaloids, steroids, tannins, phenols, glycosides and lignin were positive in all extracts, flavonoids and terpenoids were detected only in methanol and ethanol extracts which are postulated to play their role in this inhibitory activity owing to their greater solubility (Pratima and Mathad 2011). In another study, Pratima and Mathad in 2017 identified the potential antibacterial activity of the seed coat and cotyledon of *C. cajan* against *S. aureus, S. typhi and B. subtilis* and proposed that these extracts may be exploited for the development of antimicrobial and alternative remedies for infections and diseases caused by respective pathogens (Pratima and Mathad 2018). Devi et al. in 2016 reported that extracts of seed, root and leaf of *C. cajan* are antibacterial in nature. Methanol extracts from all parts of plant were effective against Gram positive bacteria, *Streptococcus* sp and *Klebsiella* sp while ethanol extracts were effective against *E. coli* and *Pseudomonas* sp. They postulated the plant extracts as the possible formulation for new antimicrobial drugs.

Kushtia region of Bangladesh houses many industries and the effluent water from these industries harbor several microorganisms including bacteria. Tannery, tobacco and sugar water wastes inhabiting microbes were isolated and characterized as the coliform (Gram negative) bacteria. Effectiveness of leaves extracts of *Cajanus cajan* L. in aqueous and ethanolic solvents was studied for inhibition of microbial growth from these waste water sources. Ethanol and ethyl acetate extract showed better antibacterial activity against all these bacterial strains (Shirina Khatun et al. 2013). In continuation to this in the year 2017, Rahman et al. showed that ethyl acetate extract of *pigeon pea* leaves had much better antibacterial activities on coliforms than the chloroform & hexane extracts and hence can be exploited against these isolated coliforms from industrial waste water.

Banala et al. in 2014 isolated aqueous as well as organic solvent (acetone, chloroform, methanol and ethanol) mediated *Cajanus* extracts from its leaf, flower, pod, stem as well as roots. The extracts showed varying degree of inhibitory effect against gram +ve and gram -ve bacterial species viz. *Staphylococcus aureus* and *Escherichia coli.* Organic extracts showed comparatively better inhibition efficiency than the aqueous extract and thus were proposed for synthesis of broad-spectrum antibacterial drugs. Among all organic solvents ethanolic and methanolic extracts stood out of the rest and to be more precise, methanolic extracts were most effective against *S. aureus* and ethanolic extracts against *E. coli*. The study conducted by Paul et al. in 2019 confirmed the efficacy of leaf extracts both against the bacterial (gram + and gram -) as well as against some fungal strains too (yeast and mold). Brito et al. 2012 have also reported that presence of flavonoids, alkaloids and tannins in pigeon pea extract has clinically established antifungal activity. This thus indicates that the plant holds great potential in the search for novel antimicrobials and it should be well studied to isolate and identify the bioactive compounds in treating the debilitating ailments caused by these pathogens (Abdulfatai et al. 2018).

### *3.6.2. Antiviral Activity*

Comparatively less research work has been done on viruses, though a couple of studies do report the anti-viral action of *cajanus cajan.* Quite recently, Prathima et al. (2017) reported that components such as tannins, alkaloids and flavonoids present in *cajanus cajan* may also contribute to its antiviral activity. Inhibition of viral titer by the hot-water extract of stem *in-vitro* as well as *in vivo* inhibition of MV replication was shown by Nwondo et al. in 2011. The *in vivo* anti-MV assay confirmed that the hot-water & ethanol extracts of pigeon pea roots as well as stems had a promising activity against MV as it lowered down the hemagglutination titer to minute levels (0.1 and 0.2). The cold-water extracts demonstrated little lesser inhibitory activity than the rest.

## 3.7. Hepatoprotective

Liver carries out several crucial functions in the body, from regulating carbohydrate, protein and fat metabolism to cleansing and excretory activities and drug detoxification. It helps in digestive activities *via* bile secretion, secretes hormones and play role in synthesis of metabolites such as plasma proteins, blood clotting factors etc. Acute and chronic liver ailments include hepatitis, cirrhosis and alcoholic liver disease. The hepatoprotective function of pigeon pea is well documented in various studies where it is reported to provide protection as well as cure against hepatitis/ jaundice and alcohol-induced liver damage too.

There are several methods to identify hepatotoxicity and to assign hepatoprotective role to plant extracts. One of the widely used marker in case of liver toxicity is assessment of biochemical Alanine aminotransferase (ALT/SGPT), Aspartate aminotransferase (AST/SGOT), cholesterol, bilirubin and glucose which are elevated in serum in case of liver damage. Drug reactions are also known to cause liver injury such as use of Acetaminophen (APAP) or paracetamol when used as a therapeutic drug for fever or pain thereby causing renal and hepatic injury in human and animals. These drugs cause aggravated oxidative reaction which in turn lead to liver damage. Thus, natural non-toxic food sources of antioxidant molecules are researched for in this aspect. Maha et al. 2014 found that proteins of Cajanus seed extract had the abilities to downregulate high levels of free radicals generated upon administering APAP or post APAP applications in test animals. The extracts were shown to have hepatoprotective effect by improving liver functions, ameliorating hepatorenal enzymes and overall reducing all signs of APAP damages including necrosis, and normalizing hepatic cells structure. This makes it hopeful for *C. cajan* proteins to be useful as a new and safe treatment for hepatorenal injuries. A polyherbal formulation (PHF) of few plant materials viz. *C. cajan* (Whole plant), *O. turpenthum* (Root), M. *pudica* (Root), *U. picta* (Root) and *L. inermis* (Leaves) in methanol was prepared to enhance the therapeutic effectiveness and improving the bioavailability of active molecules. This PHF was found to be effective as an *in vivo* antioxidant in ameliorating liver damage by elevating the otherwise reduced levels of liver cytosolic SOD and GSH (Arka et al. 2015).

In Taiwan aborigine population, “Taitung No.-3”, a new variety of pigeon pea is eaten owing to its great antioxidant and anthocyanin properties. A study was conducted by Chang et al. in 2019 to identify how the methanolic extract of *Cajanus* (MECC) lowers down the cholesterol levels. Molecular approach utilizing expression of LDLR (Low Density Lipoprotein Receptor) and its regulating protease proprotein convertase subtilisin/kexin type 9 (PCSK9) was studied in model system for hepatic studies. While LDLR resides on hepatocytes membrane, its function is to bind the plasma localized LDL-cholesterol, not considered good for human health, pinch itself in a clathrin-coated pit from surface and get into cytosol for lysosomal degradation after which LDLR recycles to join the membrane. While it is regulated post-translationally through degradation by PCSK9. The expression of LDLR and PCSK9 was monitored from cells treated with MECC. It was found that MECC drastically reduced the expression of PCSK9 by inhibiting its promoter, this led to enhanced LDLR activity. The biochemical component responsible for this cholesterol-modulating effect (LDL-C) was found to be cajaninstilbene acid which is one of the main antioxidant molecules in *Cajanus*. This outcome opens up a new avenue to explore the applications of *Cajanus* leaves as cholesterol-lowering agent (Chang et al. 2019). Methanol extract of the whole plant of *Cajanus* was observed to be hepatoprotective to rat liver cells when treated with Carbon Tetrachlorode. Intraperitoneal dose of methanol-extract cajanus treated rat had reduced toxin-mediated rise in serum levels of SGPT up to 50.22% (Ahsan et al. 2009). In addition to naturally occurring CSA in Pigeon pea, endophytic fungi *F. solani* and *F. proliferatum* have recently been reported to contribute in natural production of CSA in Pigeon pea (Zhao et al. 2012).

## Conclusion

While traditionally Pigeon pea was taken as food alone, several regions over the globe however realized the illness-curing potential of this crop over the time and used extracts of it as medicine for treatment of various ailments. Pigeon pea as we know now, harbors a wide-range of secondary metabolites having different biochemistries which makes it an ideal system for drug discovery. Being rich in antioxidants and bioactive constituents, intake of pigeon pea helps in avoiding oxidative stress and thereby avoiding related metabolic disorders. Reports suggesting antibacterial, anti-diabetic, antioxidant, anti-cancerous, hepatoprotective or many other roles of the constituents thus calls for detailed study of them. Structural, functional characterization needs to be carried out in detail to get an understanding of the mode of these biochemical. Genomics-assisted study of these phytochemicals would enable and hasten drug discovery which is a promising application. To achieve the maximum potential of *Cajanus cajan*, more research work and focus on isolating, characterizing and quantifying phytocompounds along with addressing research gaps regarding its phytochemistry need to be done (Nix, Paull and Colgrave 2015). These phytochemicals are potential raw materials which might be available as a drug component in near future for curing diseases, though research is required for quality control measures so as to realize the full potential of these components. Also, it is the right time to merge traditional medicine based research with the mainstream health care R&D to explore and avail their due use in curing diseases or in overall well-being as such. In conclusion, nutrition and health care are found to be strongly intertwined and hence plants are consumed both as food and for its medicinal usages.

## REFERENCES

Abdulfatai, K, Abdullahi B, Jaafaru I. A, Rabiu I. 2018. "Antibacterial activity of pigeon pea (*Cajanus cajan*) leaf extracts on Salmonella and Shigella Species Isolated from stool sample in patients attending baraudikkopeadiatric unit kaduna." *European Journal of Biotechnology and Bioscience 6*: 01-08.

Abo-Zeid, Mona AM, Negm S. Abdel-Samie, Ayman A. Farghaly, and Emad M. Hassan. 2018. "Flavonoid fraction of *Cajanus cajan* prohibited the mutagenic properties of cyclophosphamide in mice in vivo." *Mutation Research/Genetic Toxicology and Environmental*

*Mutagenesis* 826:1-5. doi: https://doi.org/10.1016/j.mrgentox.2017.12.004.

Ade-Omowaye, B. I. O., G. A. Tucker, and I. Smetanska. 2015. "Nutritional potential of nine underexploited legumes in Southwest Nigeria." *International Food Research Journal* 22: 798-06.url:http://www.ifrj.upm.edu.my/.

Agnese, A. M., C. Perez, and J. L. Cabrera. 2001. "Adesmiaaegiceras: antimicrobial activity and chemical study." *Phytomedicine* 8:389-94. doi: https://doi.org/10.1078/0944-7113-00059.

Ahmad, Layeeq, MdMujahid, Anuradha Mishra, and MdAzizur Rahman. 2019. "Protective role of hydroalcoholic extract of *Cajanus cajan* Linn leaves against memory impairment in sleep deprived experimental rats." *Journal of Ayurveda and Integrative Medicine*. doi:https://doi.org/10.1016/j.jaim.2018.08.003.

Ahsan, Md Rajib, Km Monirul Islam, Israt Jahan Bulbul, Md Ashik Musaddik, and Ekramul Haque. 2009. "Hepatoprotective activity of methanol extract of some medicinal plants against carbon tetrachloride-induced hepatotoxicity in rats." *Eur J Sci Res* 37: 302-10. url:http://www.eurojournals.com/ejsr.htm.

Ajaiyeoba, E. O., O. O. Ogbole, O. O. Abiodun, J. S. Ashidi, P. J. Houghton, and Colin W. Wright. 2013."Cajachalcone: An antimalarial compound from *Cajanus cajan* leaf extract." *Journal of Parasitology Research* 2013.doi:https://doi.org/10.1155/2013/703781.

Akinsulie, A. O., E. O. Temiye, A. S. Akanmu, F. E. A. Lesi, and C. O. Whyte. 2005."Clinical evaluation of extract of *Cajanus cajan* (Ciklavit®) in sickle cell anaemia." *Journal of Tropical Pediatrics* 51: 200-05.doi:https://doi.org/10.1093/tropej/fmh097.

Akojie, F. O. B., and LW-M. Fung. 1992. "Antisickling activity of hydroxybenzoic acids in *Cajanus cajan*." *Planta medica* 58: 317-20.Doi: 10.1055/s-2006-961475.

Alvarez-Mercado, José Manuel, Froylán Ibarra-Velarde, Miguel Ángel Alonso-Díaz, Yolanda Vera-Montenegro, José Guillermo Avila-Acevedo, and Ana MaríaGarcía-Bores. 2015. "In vitro antihelmintic effect of fifteen tropical plant extracts on excysted flukes of *Fasciola hepatica*." *BMC veterinary research* 11: 45.doi: https://doi.org/10.1186/s12917-015-0362-4.

Anwar, Md Mazharul, Marjina Akhter Kalpana, Bithika Bhadra, Shahnaz Rahman, Sanjoy Sarker, Majeedul H. Chowdhury, and Mohammed Rahmatullah. 2010. "Antihyperglycemic activity and brine shrimp lethality studies on methanol extract of *Cajanus cajan* (L.) Millsp. leaves and roots." *Advances in Natural and Applied Sciences* 4: 311-16. ISSN 1995-0772.

Arka, Ghosh, Kundu Anindita, Seth Ankit, Singh Anil Kumar, and Maurya Santosh Kumar. 2015."Preliminary evaluation of hepatoprotective potential of the polyherbal formulation." *Journal of Intercultural Ethnopharmacology* 4:118-124.doi:https://dx.doi.org/10.5455%2Fjice.20141121060725.

Aruna, T., and S. Devindra. 2016. "Nutritional and Anti-Nutritional Characteristics of Two Varieties of Red gram (*Cajanus cajan*, L) Seeds." *International Journal of Scientific and Research Publications* 6:567-72. url:http://ijsrp.org/.

Ashidi, J. S., P. J. Houghton, P. J. Hylands, and T. Efferth. 2010. "Ethnobotanical survey and cytotoxicity testing of plants of South-western Nigeria used to treat cancer, with isolation of cytotoxic constituents from *Cajanus cajan* Millsp. leaves." *Journal of Ethnopharmacology* 128: 501-12.doi:https://doi.org/10.1016/j.jep.2010.01.009.

Banala, Mahitha, Srikanth Kagithoju, Archana Pamulaparthi, Chamundeshwari Edara and Ramaswamy Nanna. 2015. "*In vitro* antibacterial activity of leaf, seed, root, pod and flower extracts of *Cajanus cajan* (l.) millsp." *Int J Pharm Pharm Sci* 7:265-68.url:https://www.researchgate.net/publication/280326884_In_vitro_antibacterial_activity_of_leaf_seed_root_pod_and_flower_extracts_of_Cajanus_Cajan_L_millsp.

Baskar, Venkidasamy, Se Won Park, and Shivraj Hariram Nile. 2016. "An update on potential perspectives of glucosinolates on protection against microbial pathogens and endocrine dysfunctions in humans." *Critical reviews in food science and nutrition* 56: 2231-49.doi:https://doi.org/ 10.1080/10408398.2014.910748.

Bhanumati, S., S. C. Chhabra, and S. R. Gupta. 1979. "Cajaisoflavone, a new prenylated isoflavone from *Cajanus cajan*." *Phytochemistry* 18: 1254.

Bhattacharjee, S., and G. D. Sharma. 2015."Effect of Arbuscularmyco-rrhizal fungi (AM fungi) and Rhizobium on the nutrient uptake of pigeon pea plant." *Int J Adv Res* 3: 833-36.

Bhattacharjee, Saraj., A. K., Manoj Kumar, and S. K. Sharma. 2013. "Phosphorus, sulfur and cobalt fertilization effect on yield and quality of soybean (Glycine max L. Merrill) in acidic soil of northeast India." *Indian J Hill Farm* 26: 63-66.url:http://www.kiran.nic.in/.

Boye, Joyce, Fatemeh Zare, and Alison Pletch. 2010. "Pulse proteins: Processing, characterization, functional properties and applications in food and feed." *Food research international* 43: 414-31.doi:https:// doi.org/10.1016/j.foodres.2009.09.003.

Bramati, Lorenzo, Francesca Aquilano, and Piergiorgio Pietta. 2003. "Unfermented rooibos tea: quantitative characterization of flavonoids by HPLC− UV and determination of the total antioxidant activity." *Journal of agricultural and food chemistry* 51:7472-74.doi:https:// doi.org/10.1021/jf0347721.

Brito, Samara A., Fabíola FG Rodrigues, Adriana R. Campos, and José GM Da Costa. 2012. "Evaluation of the antifungal activity and modulation between *Cajanus cajan* (L.) Millsp. leaves and roots ethanolic extracts and conventional antifungals." *Pharmacognosy magazine* 8: 103-06.doi:https://dx.doi.org/10.4103%2F0973-1296.96550.

Cai, Jia-Zhong, Rong Tang, Gui-Fu Ye, Sheng-Xiang Qiu, Nen-Ling Zhang, Ying-Jie Hu, and Xiao-Ling Shen. 2015. "A halogen-containing stilbene derivative from the leaves of *Cajanus cajan* that induces osteogenic differentiation of human mesenchymal stem cells." *Molecules* 20: 10839-47. doi: https://doi.org/10.3390/molecules 200610839.

Campos-Vega, Rocio, B. Dave Oomah, Guadalupe Loarca-Piña, and Haydé Azeneth Vergara-Castañeda. 2013. "Common beans and their non-digestible fraction: cancer inhibitory activity—an overview." *Foods* 2: 374-92.doi:https://doi.org/10.3390/foods2030374.

Chang, Heng-Yuan, Jia-Ru Wu, Wan-Yun Gao, Huei-Ru Lin, Pei-Yi Chen, Chen-I. Chen, Ming-Jiuan Wu, and Jui-Hung Yen. 2019. "The Cholesterol-Modulating Effect of Methanol Extract of Pigeon Pea (*Cajanus cajan* (L.) Millsp.) Leaves on Regulating LDLR and PCSK9 Expression in HepG2 Cells." *Molecules* 24: 493.doi:https://doi.org/10.3390/molecules24030493.

Chen, Shi-Lin, Hua Yu, Hong-Mei Luo, Qiong Wu, Chun-Fang Li, and André Steinmetz. 2016."Conservation and sustainable use of medicinal plants: problems, progress, and prospects." *Chinese medicine* 11: 37.doi:https://doi.org/10.1186/s13020-016-0108-7.

Chen, Wen-Zhang, Ling-Ling Fan, Hai-Tao Xiao, Ying Zhou, Wan Xiao, Jian-Ta Wang, and Lei Tang. 2014. "First total synthesis of natural products cajanolactone A and cajanonic acid A." *Chinese Chemical Letters* 25: 749-51.doi:https://doi.org/10.1016/j.cclet.2014.03.027.

Chidebelu, Paul, Ifeoma Onovo, and Emeka I. Nweze. 2019." Antimicrobial activities of *Cajanus cajan* L., Phaseolus vulgaris L. and Vignaungui-culata L. against some bacterial and fungal isolates." *Journal of Basic Pharmacology and Toxicology* 3:29-33. doi: http://www.scigreen.com/index.php/JBPT/article/view/58.

Constantinou, Andreas I., N. Kamath, and J. S. Murley. 1998. "Genistein inactivates bcl-2, delays the G2/M phase of the cell cycle, and induces apoptosis of human breast adenocarcinoma MCF-7 cells." *European journal of cancer* 34: 1927-34. doi: https://doi.org/10.1016/S0959-8049(98)00198-1.

Cooksey, Christopher J., Jagroop S. Dahiya, Peter J. Garratt, and Richard N. Strange. 1980. "Two novel stilbene-2-carboxylic acid phytoalexins from *Cajanus cajan*." *Phytochemistry* 21: 2935-38. doi: https://doi.org/10.1016/0031-9422(80)85072-2.

Cooksey, Christopher J., Peter J. Garratt, Jagroop S. Dahiya, and Richard N. Strange. 1983. "Sucrose: a constitutive elicitor of phytoalexin synthesis." *Science* 220: 1398-1400.Doi: 10.1126/science.220.4604. 1398.

Dahiya, Jagroop S. 1987."Reversed-phase high-performance liquid chromatography of *Cajanus cajan* phytoalexins." *Journal of Chromatography A* 409: 355-59.doi:https://doi.org/10.1016/S0021-9673(01)86812-6.

Dahiya, Jagroop S. 1991."Cajaflavanone and cajanone released from *Cajanus cajan* (L. Mill sp.) roots induce nod genes of Bradyrhizobium sp." *Plant and soil* 134: 297-04.doi:https://doi.org/10.1007/BF00012049.

Dahiya, Jagroop S., Richard N. Strange, Kevin G. Bilyard, Christopher J. Cooksey, and Peter J. Garratt. 1984."Two isoprenylated isoflavone phytoalexins from *Cajanus cajan.*" *Phytochemistry* 23: 871-73.doi:https://doi.org/10.1016/S0031-9422(00)85046-3.

Das, Subhasish, K. Charan Teja, Sandip Mukherjee, Soma Seal, Rajesh Kumar Sah, Buddhadeb Duary, Ki-Hyun Kim, and Satya Sundar Bhattacharya. 2018."Impact of edaphic factors and nutrient management on the hepatoprotective efficiency of Carlinoside purified from pigeon pea leaves: an evaluation of UGT1A1 activity in hepatitis induced organelles." *Environmental research* 161: 512-23. doi: https://doi.org/10.1016/j.envres.2017.11.054.

Devi, R. Raveena, R. Premalatha, and A. Saranya. 2016. "Comparative analysis of phytochemical constituents and antibacterial activity of leaf, seed and root extract of *Cajanus cajan* (L.) Mill sp." *International Journal of Current Microbiology and Applied Sciences* 5: 485-94.doi:http://dx.doi.org/10.20546/ijcmas.2016.503.057.

Dolui, A. K., and R. Segupta. 2012. "Antihyperglycemic effect of different solvent extracts of leaves of *Cajanus cajan* HPLC profile of the active extracts." *Asian J Pharm Clinical Res* 5: 116-9.

Duke, J. A., and R. M. Polhill. 1981. "Seedlings of leguminosae." *Advances in legume systematics*.

Duke, James. 2012. *Handbook of legumes of world economic importance*. Springer Science & Business Media.

Duker-Eshun, George, Jerzy W. Jaroszewski, William A. Asomaning, Francis Oppong-Boachie, and S. Brøgger Christensen. 2004. "Antiplasmodial constituents of *Cajanus cajan.*" *Phytotherapy Research: An International Journal Devoted to Pharmacological and Toxicological Evaluation of Natural Product Derivatives* 18: 128-30.doi: https://doi.org/10.1002/ptr.1375.

Ekeke, G. I., and F. O. Shode. 1985."The reversion of sickled cells by *Cajanus cajan.*" *Planta medica* 51: 504-07.Doi: 10.1055/s-2007-969576.

Ekeke, G. I., and F. O. Shode. 1990. "Phenylalanine is the predominant antisickling agent in *Cajanus cajan* seed extract." *Planta medica* 56: 41-43.doi: 10.1055/s-2006-960880.

El-Adawy, T. A., E. H. Rahma, A. A. El-Bedawey, and A. E. El-Beltagy. 2003."Nutritional potential and functional properties of germinated mung bean, pea and lentil seeds." *Plant Foods for Human Nutrition* 58: 1-13.doi:https://doi.org/10.1023/B:QUAL.0000040339.48521.75.

El-Mawla, Abd., Ahmed MA, and Husam Eldien H. Osman. 2011."HPLC analysis and role of the Saudi Arabian propolis in improving the pathological changes of kidney treated with monosodium glutamate." *Spatula DD* 1:119-27. doi: 10.5455/spatula.20110803105445.

Ezike, Adaobi C., Peter A. Akah, Charles C. Okoli, and Chinwe B. Okpala. 2010. "Experimental evidence for the antidiabetic activity of *Cajanus cajan* leaves in rats." *Journal of basic and clinical pharmacy* 1: 81-84.

Fagbohun Emmanuel, D., M. David Oluwole, I. Adeyeye Emmanuel, and O. Oyedele. 2010."Chemical composition and antibacterial activities of some selected traditional medicinal plants used in the treatment of gastrointestinal infections in Nigeria." *International Journal of Pharmaceutical Sciences Review and Research* 5:192-97. url: https://www.researchgate.net/publication/274715713_Chemical_composition_and_antibacterial_activities_of_some_selected_traditional_medicinal_plants_used_in_the_treatment_of_gastrointestinal_infections_in_Nigeria.

Fu, Yujie, OnatKadioglu, Benjamin Wiench, Zuofu Wei, Chang Gao, Meng Luo, Chengbo Gu, Yuangang Zu, and Thomas Efferth.

2015."Cell cycle arrest and induction of apoptosis by cajanin stilbene acid from *Cajanus cajan* in breast cancer cells." *Phytomedicine* 22: 462-68.doi:https://doi.org/10.1016/j.phymed.2015.02.005.

Fu, Yujie, OnatKadioglu, Benjamin Wiench, Zuofu Wei, Wei Wang, Meng Luo, Xiaohe Yang, Chengbo Gu, Yuangang Zu, and Thomas Efferth.2015. "Activity of the antiestrogenic cajanin stilbene acid towards breast cancer." *The Journal of Nutritional Biochemistry* 26: 1273-82.doi:https://doi.org/10.1016/j.jnutbio.2015.06.004.

Fu, Yu-Jie, Wei Liu, Yuan-Gang Zu, Mei-Hong Tong, Shuang-Ming Li, Ming-Ming Yan, Thomas Efferth, and Hao Luo. 2008. "Enzyme assisted extraction of luteolin and apigenin from pigeon pea [*Cajanus cajan* (L.) Millsp.] leaves." *Food Chemistry* 111: 508-12. doi: https://doi.org/10.1016/j.foodchem.2008.04.003.

Fu, Yujie, Yuangang Zu, Wei Liu, ChunlianHou, Liyan Chen, Shuangming Li, Xiaoguang Shi, and Meihong Tong. 2007."Preparative separation of vitexin and isovitexin from pigeonpea extracts with macroporous resins." *Journal of Chromatography A* 1139: 206-13. doi: https://doi.org/10.1016/j.chroma.2006.11.015.

Fu, Yujie, Yuangang Zu, Wei Liu, Lin Zhang, Meihong Tong, Thomas Efferth, Yu Kong, ChunlianHou, and Liyan Chen. 2008." Determination of vitexin and isovitexin in pigeon pea using ultrasonic extraction followed by LC-MS." *Journal of separation science* 31: 268-75.doi:https://doi.org/10.1002/jssc.200700312.

Ganga Raju, M., and K. Kumara Swamy. 2018. "Anti-inflammatory and antiradical potential of methanolic extract of *Cajanus cajan*." *Asian Journal of Pharmacy and Pharmacology* 4:860-64. doi: https://doi.org/10.31024/ajpp.2018.4.6.21.

Gao, Yuan, Jintong Zhao, Yuangang Zu, Yujie Fu, Lu Liang, Meng Luo, Wei Wang, and Thomas Efferth. 2012. "Antioxidant properties, superoxide dismutase and glutathione reductase activities in HepG2 cells with a fungal endophyte producing apigenin from pigeon pea [*Cajanus cajan* (L.) Millsp.]." *Food research international* 49:147-52.doi:https://doi.org/10.1016/j.foodres.2012.08.001.

Green, Paul WC, Philip C. Stevenson, Monique SJ Simmonds, and Hari C. Sharma. 2003."Phenolic compounds on the pod-surface of pigeon pea, *Cajanus cajan*, mediate feeding behavior of *Helicoverpa armigera* larvae." *Journal of Chemical Ecology* 29: 811-21.doi:https://doi.org/10.1023/A:1022971430463.

Grover, J. K., S. Yadav, and V. Vats. 2002. "Medicinal plants of India with anti-diabetic potential." *Journal of ethnopharmacology* 81: 81-100.doi:https://doi.org/10.1016/S0378-8741(02)00059-4.

Grover, J. K., V. Vats, S. S. Rathi, and R. Dawar. 2001."Traditional Indian anti-diabetic plants attenuate progression of renal damage in streptozotocin induced diabetic mice." *Journal of Ethnopharmacology* 76: 233-38.doi:https://doi.org/10.1016/S0378-8741(01)00246-X.

Gu, C. B., H. Ma, W. J. Ning, L. L. Niu, H. Y. Han, X. H. Yuan, and Y. J. Fu. 2018. "Characterization, culture medium optimization and antioxidant activity of an endophytic vitexin-producing fungus *Dichotomopilus funicola* Y3 from pigeon pea [*Cajanus cajan* (L.) Millsp.]." *Journal of applied microbiology* 125: 1054-65.doi:https://doi.org/10.1111/jam.13928.

Gui-Yun, W. U., Xiao Zhang, G. U. O. Xue-Ying, H. U. O. Lu-Qiong, L. I. U. Hong-Xin, S. H. E. N. Xiao-Ling, Q. I. U. Sheng-Xiang, H. U. Ying-Jie, and T. A. N. Hai-Bo. 2019. "Prenylated stilbenes and flavonoids from the leaves of *Cajanus cajan*." *Chinese journal of natural medicines* 17: 381-86.doi:https://doi.org/10.1016/S1875-5364(19)30044-5.

Halliwell, Barry. 1989."Free radicals, reactive oxygen species and human disease: a critical evaluation with special reference to atherosclerosis." *British journal of experimental pathology* 70: 737-57.PMCID: PMC2040729.PMID: 2557883.

Hassan, Emad M., Azza A. Matloub, Mona E. Aboutabl, Nabaweya A. Ibrahim, and Samy M. Mohamed. 2016. "Assessment of anti-inflammatory, antinociceptive, immunomodulatory, and antioxidant activities of *Cajanus cajan* L. seeds cultivated in Egypt and its phytochemical composition." *Pharmaceutical biology* 54:1380-91. doi: https://doi.org/10.3109/13880209.2015.1078383.

Hopp, David C., and Wayne D. Inman. "Compositions containing hypoglycemically active still benoids." U.S. Patent 6,410,596, issued June 25, 2002.

Hosseinpour-Niazi, S., P. Mirmiran, M. Hedayati, and F. Azizi. 2015. "Substitution of red meat with legumes in the therapeutic lifestyle change diet based on dietary advice improves cardiometabolic risk factors in overweight type 2 diabetes patients: a cross-over randomized clinical trial." *European journal of clinical nutrition* 69:592-97.doi:https://doi.org/10.1038/ejcn.2014.228.

Huang, Cheng, Yang Yang, Wan-Xia Li, Xiao-Qin Wu, Xiao-Feng Li, Tao-Tao Ma, Lei Zhang, Xiao-Ming Meng, and Jun Li. 2015."Hyperin attenuates inflammation by activating PPAR-γ in mice with acute liver injury (ALI) and LPS-induced RAW264. 7 cells." *International immunopharmacology* 29: 440-47. doi: https://doi.org/10.1016/j.intimp. 2015.10.017.

Huang, Mei-Yan, Jing Lin, Kuo Lu, Hong-Gui Xu, Zhi-ZhongGeng, Ping-Hua Sun, and Wei-Min Chen. 2016. "Anti-inflammatory effects of cajan in stilbene acid and its derivatives." *Journal of agricultural and food chemistry* 64:2893-2900. doi: https://doi.org/10.1021/acs.jafc.6b00227.

Iacopini, P., M. Baldi, P. Storchi, and L. Sebastiani. 2008."Catechin, epicatechin, quercetin, rutin and resveratrol in red grape: Content, in vitro antioxidant activity and interactions." *Journal of Food Composition and Analysis* 21: 589-98.doi:https://doi.org/10.1016/j.jfca.2008.03.011.

Ingham, John L. 1976."Induced isoflavonoids from fungus-infected stems of pigeon pea (*Cajanus cajan*)." *ZeitschriftfürNaturforschung C* 31: 504-08.doi:https://doi.org/10.1515/znc-1976-9-1005.

Inman, Wayne D., and David C. Hopp. 2003. *Methods of using compositions containing hypotriglyceridemically active stilbenoids.* U.S. Patent 6,541,522, issued April 1, 2003.

Inman, Wayne D., and David C. Hopp. 2003. *Compositions containing hypoglycemically active stilbenoids.* U.S. Patent 6,552,085, issued April 22, 2003.

Iwu, M. M., A. O. Igboko, H. Onwubiko, and U. E. Ndu. 1988."Effect of cajaminose from *Cajanus cajan* on gelation and oxygen affinity of sickle cell haemoglobin." *Journal of ethnopharmacology* 23: 99-104.doi:https://doi.org/10.1016/0378-8741(88)90118-3.

Iwu, M. M., O. A. Igboko, H. Onwubiko, and U. E. Ndu. 1986. "Anti-sickling Properties of *Cajanus cajan*: Effect on Haemoglobin Gelation and Oxygen Affinity." *Planta medica* 52: 431.

Jacob, Bindu, and R. T. Narendhirakannan. 2019."Role of medicinal plants in the management of diabetes mellitus: a review." *3 Biotech* 9: 4.doi:https://doi.org/10.1007/s13205-018-1528-0.

Jeevarathinam, G., and V. Chelladurai. 2020. "Pigeon Pea." In *Pulses*, pp. 275-96. Springer, Cham. doi: https://doi.org/10.1007/978-3-030-41376-7_15.

Jiao, Jiao, Qing-Yan Gai, Xin Wang, Jing Liu, Yao Lu, Zi-Ying Wang, Xiao-Jie Xu, and Yu-Jie Fu. 2020."Effective Production of Phenolic Compounds with Health Benefits in Pigeon Pea [*Cajanus cajan* (L.) Millsp.] Hairy Root Cultures." *Journal of Agricultural and Food Chemistry* 68: 8350-61.doi:https://doi.org/10.1021/acs.jafc.0c02600.

Kang, Jihee Lee, Hye Won Lee, Hui Su Lee, In Soon Pack, Vincent Castranova, and Younsuck Koh. 2003."Time course for inhibition of lipopolysaccharide-induced lung injury by genistein: relationship to alteration in nuclear factor-κB activity and inflammatory agents." *Critical care medicine* 31: 517-24.doi: 10.1097/01.CCM. 0000049941.84695.BA.

Kaul, Sanjana, Suruchi Gupta, Maroof Ahmed, and Manoj K. Dhar. "Endophytic fungi from medicinal plants: a treasure hunt for bioactive metabolites." *Phytochemistry reviews* 11: 487-05. doi: https://doi.org/ 10.1007/s11101-012-9260-6.

Keshav, Kamaliya B. 2015. "Optimization of High Fiber Bun Formula and its Nutritional Evaluation." *International Journal of Food, Nutrition and Dietetics* 3: 89-93. doi: https://dx.doi.org/10.21088/ijfnd.2322. 0775.3315.1.

Khatun, Mst Shirina, Md Ruhul Amin, Md Manirujjaman, Md Monirul Islam, and Hossain Md Faruque. *Antibacterial activity of Cajanus cajan leaves against industrial waste water bacteria collected from Kushtia, Bangladesh.*

Kong, Yu, Yu-Jie Fu, Yuan-Gang Zu, Fang-Rong Chang, Yung-Husan Chen, Xiao-Lei Liu, Johannes Stelten, and Hans-Martin Schiebel. 2010. "Cajanuslactone, a new coumarin with anti-bacterial activity from pigeon pea [*Cajanus cajan* (L.) Millsp.] leaves." *Food chemistry* 121: 1150-55.doi:https://doi.org/10.1016/j.foodchem.2010.01.062.

Kong, Yu, Yu-Jie Fu, Yuan-Gang Zu, Wei Liu, Wei Wang, Xin Hua, and Mei Yang. 2009. "Ethanol modified supercritical fluid extraction and antioxidant activity of cajaninstilbene acid and pinostrobin from pigeon pea [*Cajanus cajan* (L.) Millsp.] leaves." *Food Chemistry* 117: 152-59.doi:https://doi.org/10.1016/j.foodchem.2009.03.091.

Kong, Yu, Zuo-Fu Wei, Yu-Jie Fu, Cheng-Bo Gu, Chun-Jian Zhao, Xiao-Hui Yao, and Thomas Efferth.2011. "Negative-pressure cavitation extraction of cajaninstilbene acid and pinostrobin from pigeon pea [*Cajanus cajan* (L.) Millsp.] leaves and evaluation of antioxidant activity." *Food chemistry* 128: 596-605.doi:https://doi.org/10.1016/j.foodchem.2011.02.079.

Kooti, Wesam, Maryam Farokhipour, Zahra Asadzadeh, Damoon Ashtary-Larky, and Majid Asadi-Samani. 2016. "The role of medicinal plants in the treatment of diabetes: a systematic review." *Electronic physician* 8: 1832.doi:https://dx.doi.org/10.19082%2F1832.

Lai, Yi-Syuan, Wei-Hsuan Hsu, Jan-Jeng Huang, and She-Ching Wu. 2012. "Antioxidant and anti-inflammatory effects of pigeon pea (*Cajanus cajan* L.) extracts on hydrogen peroxide-and lipopolysaccharide-treated RAW264. 7 macrophages." *Food & function* 3: 1294-1301.doi:https://doi.org/10.1039/C2FO30120B.

Leelaprakash, G., and S. Mohan Dass. 2011. "Invitro anti-inflammatory activity of methanol extract of Enicostem maaxillare." *International Journal of Drug Development and Research* 3: 189-96.url: http://www.ijddr.in/.

Liu, Wei, Su Zhang, Yuan-Gang Zu, Yu-Jie Fu, Wei Ma, Dong-Yang Zhang, Yu Kong, and Xiao-Juan Li. 2010. "Preliminary enrichment and separation of genistein and apigenin from extracts of pigeon pea roots by macroporous resins." *Bioresource Technology* 101:4667-75.doi:https://doi.org/10.1016/j.biortech.2010.01.058.

Liu, Wei, Yu Kong, Yuangang Zu, Yujie Fu, Meng Luo, Lin Zhang, and Ji Li. 2010. "Determination and quantification of active phenolic compounds in pigeon pea leaves and its medicinal product using liquid chromatography–tandem mass spectrometry." *Journal of Chromatography A* 1217:4723-31.doi:https://doi.org/10.1016/j.chroma.2010.05.020.

Liu, Wei, Yuan-Gang Zu, Yu-Jie Fu, Yu Kong, Wei Ma, Mei Yang, Ji Li, and Nan Wu. 2010. "Variation in contents of phenolic compounds during growth and post-harvest storage of pigeon pea seedlings." *Food chemistry* 121:732-39. doi: https://doi.org/10.1016/j.foodchem.2010.01.023.

Liu, Xiao-lei, Xin-jian Zhang, Yu-jie Fu, Yuan-gang Zu, Nan Wu, Lu Liang, and Thomas Efferth. 2011. "Cajanol inhibits the growth of Escherichia coli and Staphylococcus aureus by acting on membrane and DNA damage." *Planta medica* 77:158-63. doi: http://dx.doi.org/10.1055/s-0030-1250146.

Luo, Meng, Xia Liu, Yuangang Zu, Yujie Fu, Su Zhang, Liping Yao, and Thomas Efferth. 2010."Cajanol, a novel anticancer agent from Pigeon pea [*Cajanus cajan* (L.) Millsp.] roots, induces apoptosis in human breast cancer cells through a ROS-mediated mitochondrial pathway." *Chemico-Biological Interactions* 188: 151-60. doi: https://doi.org/10.1016/j.cbi.2010.07.009.

Luo, Qing-Feng, Lan Sun, Jian-Yong Si, and Di-Hua Chen. 2008. "Hypocholesterolemic effect of stilbenes containing extract-fraction from *Cajanus cajan* L. on diet-induced hypercholesterolemia inmice." *Phytomedicine* 15:932-39.doi:https://doi.org/10.1016/j.phymed.2008.03.002.

Mahitha, B., P. Archana, Md H. Ebrahimzadeh, K. Srikanth, M. Rajinikanth, and N. Ramaswamy. 2015. "In vitro antioxidant and pharmacognostic studies of leaf extracts of *Cajanus cajan* (l.) millsp." *Indian Journal of Pharmaceutical Sciences* 77: 170-77.doi: https://doi.org/10.4103/0250-474x.156555.

Marathe, Sushama A., V. Rajalakshmi, Sahayog N. Jamdar, and Arun Sharma. 2011. "Comparative study on antioxidant activity of different varieties of commonly consumed legumes in India." *Food and Chemical Toxicology* 49:2005-12. doi: https://doi.org/10.1016/j.fct.2011.04.039.

Maritim, A. C., RA Sanders, and J. B. Watkins III. 2003."Diabetes, oxidative stress, and antioxidants: a review." *Journal of biochemical and molecular toxicology* 17: 24-38.doi:https://doi.org/10.1002/jbt.10058.

Mathew, Deepu, T. M. Manila, P. Divyasree, and Sandhya Rajan VTK. 2017."Therapeutic molecules for multiple human diseases identified from pigeon pea (*Cajanus cajan* L. Millsp.) through GC–MS and molecular docking." *Food Science and Human Wellness* 6: 202-16.doi:https://doi.org/10.1016/j.fshw.2017.09.003.

Mathukia, R. K., H. P. Ponkia, and A. M. Polara. "In situ conservation and zinc fertilization for rain fed pigeon pea (*Cajanus cajan* L.)." *The Bioscan* 11: 247-50.

Menichini, Federica, Monica R. Loizzo, Marco Bonesi, FilomenaConforti, Damiano De Luca, Giancarlo A. Statti, Bruno de Cindio, Francesco Menichini, and Rosa Tundis. 2011."Phytochemical profile, antioxidant, anti-inflammatory and hypoglycemic potential of hydroalcoholic extracts from *Citrus medica* L. cv Diamante flowers, leaves and fruits at two maturity stages." *Food and Chemical Toxicology* 49:1549-55.doi:https://doi.org/10.1016/j.fct.2011.03.048.

Muangman, Thanchanok, Wichet Leelamanit, and Prapaipat Klungsupaya. 2011. "Crude proteins from pigeon pea (*Cajanus cajan* (L.) Millsp) possess potent SOD-like activity and genoprotective effect against H2O2 in TK6 cells." *Journal of Medicinal Plants Research* 5: 6977-86.doi:https://doi.org/10.5897/JMPR11.996.

Mueller, Monika, Stefanie Hobiger, and AloisJungbauer. 2010."Anti-inflammatory activity of extracts from fruits, herbs and spices." *Food Chemistry* 122:987-96.doi:https://doi.org/10.1016/j.foodchem.2010.03.041.

Nahar, Laizuman, FatemaNasrin, RonokZahan, AnamulHaque, EkramulHaque, and Ashik Mosaddik. 2014. "Comparative study of antidiabetic activity of *Cajanus cajan* and *Tamarin dusindica* in alloxan-induced diabetic mice with a reference to in vitro antioxidant activity." *Pharmacognosy research* 6: 180-87.doi: https://dx.doi.org/10.4103%2F0974-8490.129043.

Nix, Aaron, Cate A. Paull, and Michelle Colgrave. 2015. "The flavonoid profile of pigeonpea, *Cajanus cajan*: a review." *SpringerPlus* 4: 125.doi:https://doi.org/10.1186/s40064-015-0906-x.

Nwodo, U. U., A. A. Ngene, C. U. Iroegbu, O. A. L. Onyedikachi, V. N. Chigor, and A. I. Okoh. 2011. "In vivo evaluation of the antiviral activity of *Cajanus cajan* on measles virus." *Archives of virology* 156: 1551-57.Doi: 10.1007/s00705-011-1032-x.

Okigbo, R. N., and O. D. Omodamiro. 2007."Antimicrobial effect of leaf extracts of pigeon pea (*Cajanus cajan* (L.) Millsp.) on some human pathogens." *Journal of herbs, spices & medicinal plants* 12: 117-27.doi:https://doi.org/10.1300/J044v12n01_11.

Onah, Ogoda., Johnson, Paul I. Akubue, and George B. Okide. 2002. "The kinetics of reversal of pre-sickled erythrocytes by the aqueous extract of *Cajanus cajan* seeds." *Phytotherapy Research: An International Journal Devoted to Pharmacological and Toxicological Evaluation of Natural Product Derivatives* 16: 748-50.doi: https://doi.org/10.1002/ptr.1026.

Orni, Proma Roy, Sazzad Zaman Ahmed, MarzanaMonefa, Tanzim Khan, and PriteshRanjan Dash. 2018. "Pharmacological and phytochemical properties of *Cajanus cajan* (L.) Huth. (Fabaceae): A review." *International Journal of Pharmaceutical Science and Research* 3: 27-37.

Pal, Dilipkumar, Pragya Mishra, NeetuSachan, and Ashoke K. Ghosh. 2011."Biological activities and medicinal properties of *Cajanus cajan* (L) Millsp." *Journal of advanced pharmaceutical technology & research* 2: 207-14.doi:https://dx.doi.org/10.4103%2F2231-4040.90874.

Palanisamy, Uma, Thamilvaani Manaharan, Ling Lai Teng, Ammu KC Radhakrishnan, Thavamanithevi Subramaniam, and Theanmalar Masilamani. 2011. "Rambutan rind in the management of hyperglycemia." *Food Research International* 44: 2278-82.doi: https://doi.org/10.1016/j.foodres.2011.01.048.

Patel, Neeraj K., and Kamlesh K. Bhutani. 2014. "Pinostrobin and Cajanus lactone isolated from *Cajanus cajan* (L.) leaves inhibits TNF-α and IL-1β production: In vitro and in vivo experimentation." *Phytomedicine* 21:946-53.doi:https://doi.org/10.1016/j.phymed.2014.02.011.

*Patent Application Number: CN101485714A.*

Patil, B. S., and V. S. Mastiholimath. 2011."Wound healing activity of hydrogel obtained from pigeon pea (*Cajanus cajan*) seed husk." *J Chem Phar Sci* 4: 108-10.

Picerno, Patrizia, Teresa Mencherini, Maria Rosaria Lauro, Francesco Barbato, and Rita Aquino. 2003. "Phenolic constituents and antioxidant properties of Xanthosomaviolaceum leaves." *Journal of agricultural and food chemistry* 51:6423-28.doi:https://doi.org/10.1021/jf030284h.

Prabhakar, M. C., HassinaBano, I. Kumar, M. A. Shamsi, and S. Y. Khan. 1981."Pharmacological investigations on vitexin." *Planta medica* 43: 396-403. DOI: 10.1055/s-2007-971532.

Pratima, H., and P. Mathad. 2018."*In vitro* antibacterial activities of different solvent extracts of *cajanus cajan* L. Seed coat and cotyledon." *Asian Journal of Pharmaceutical and Clinical Research 11*:325-3-28. url: http://gukir.inflibnet.ac.in/handle/123456789/10.22159/ajpcr.2018.v11i5.23610.

Pratima, H., and Pratima Mathad. 2011."Antibacterial activity of various leaf extract of *Cajanus cajan* L." *Bioscan* 6: 111-14.

Pratima, H., and Pratima Mathad. 2017. "Comparative Study On Pharmacognostic and Phytochemical Composition of Seed Coat And Cotyledon Of *Cajanus cajan* L." *International Journal of Pharmaceutical Sciences and Research* 8: 1751-57. doi:10.13040/IJPSR.0975-8232.8(4).1751-57.

Preston, Nigel W. "Cajanone: an antifungal isoflavanone from *Cajanus cajan*." *Phytochemistry (UK)* (1977).

Rahman Md. Habibur., Md. Mizanur Rahman, Methun Chakraborti, Feroja Easmin, Rajib Majumdar, Sabrina Jahan. 2016. *International journal of Innovative Pharmaceutical Sciences and Research* 4:605-17. ISSN (online) 2347-2154.

Rahman, Md Motiar, Md Sahab Uddin, Md Rashed Nejum, S. M. Al Din, and G. S. Uddin. 2017."Study on antibacterial activity of *Cajanus cajan* L. against coliforms isolated from industrial waste water in Bangladesh." *Plant* 5: 13-18. Doi: 10.11648/j.plant.s.2017050501.12.

Rani, Savita, Gagan Poswal, Rajesh Yadav, and M. K. Deen. 2014."Screening of pigeon pea (*Cajanus cajan* L.) seeds for study of their flavonoids, total phenolic content and antioxidant properties." *International Journal of Pharmaceutical Sciences Review and Research* 28: 90-94. url:http://www.globalresearchonline.net/.

Rizk, Maha Z., Hanan F. Aly, Dina M. Abo-Elmatty, M. M. Desoky, N. Ibrahim, and Eman A. Younis. 2016. "Hepatoprotective effect of Caesal piniagilliesii and *Cajanus cajan* proteins against acetoaminophen overdose-induced hepatic damage." *Toxicology and industrial health* 32: 877-907. doi: https://doi.org/10.1177%2F0748233713503030.

Sarkar, Rhitajit, Bibhabasu Hazra, Sourav Mandal, Santanu Biswas, and Nripendranath Mandal. 2009."Assessment of in vitro antioxidant and free radical scavenging activity of *Cajanus cajan*." *Journal of Complementary and Integrative Medicine* 6.doi:https://doi.org/10.2202/1553-3840.1248.

Sarkar, S., S. Panda, K. Yadav, and P. Kandasamy. 2018. "Pigeon pea (*Cajanus cajan*) an important food legume in Indian scenario-A review." *Legum Res* 402: 1-10. Doi:10.18805/LR-4021.

Saxena, Rachit K., K. B. Saxena, and Rajeev K. Varshney. 2019. "Pigeon pea (*Cajanus cajan* L. Millsp.): An Ideal Crop for Sustainable Agriculture." In *Advances in Plant Breeding Strategies: Legumes*. Springer, Cham. 409-29.doi:https://doi.org/10.1007/978-3-030-23400-3_11.

Schuster, Roswitha, Wolfgang Holzer, Hannes Doerfler, Wolfram Weckwerth, Helmut Viernstein, Siriporn Okonogi, and Monika Mueller. 2016."*Cajanus cajan*–a source of PPARγ activators leading to anti-inflammatory and cytotoxic effects." *Food & function* 7: 3798-3806.doi:https://doi.org/10.1039/C6FO00689B.

Sharma, Sheel., Nidhi Agarwal, and Preeti Verma. 2011."Pigeon pea (*Cajanus cajan* L.): a hidden treasure of regime nutrition." *Journal of Functional and Environmental Botany* 1: 91-101.

Siddhartha, Singh., Mehta, Archana., John Jinu and Mehta, Pradeep. 2009."Anthelmintic Potential of Andrograph ispaniculata, *Cajanus cajan* and Silybum marianum." *Pharmacognosy Journal* 1:243-45.

Singh, Anupma., S. K., and A. Vaishampayan. 2017. "Assessment of genetic diversity and symbiotic effectiveness of Pigeon pea (*Cajanus cajan*) nodulating bradyrhizobia." *Indian Res J Genet Biotech* 9: 179-86.

Singh, Jagdish and Partha Sarathi Basu. 2012. "Non-nutritive bioactive compounds in pulses and their impact on human health: an overview." *Food and Nutrition Sciences 3*.url:http://www.scirp.org/journal/PaperInformation.aspx?PaperID=25289.

Sosa, Silvio., M. J. Balick., R. Arvigo., R. G. Esposito., C. Pizza, G. Altinier, and Aurelia Tubaro. 2002. "Screening of the topical anti-inflammatory activity of some Central American plants." *Journal of Ethnopharmacology* 81: 211-15.doi:https://doi.org/10.1016/S0378-8741(02)00080-6.

Sugishita, Etsuko, Sakae Amagaya, and Yukio Ogihara. 1981."Anti-inflammatory testing methods: comparative evaluation of mice and rats." *Journal of Pharmacobio-dynamics* 4: 565-75.doi: https://doi.org/10.1248/bpb1978.4.565.

Syed, Rabia, and Ying Wu. 2018."A review article on health benefits of Pigeon pea (*Cajanus cajan* (L.) Millsp)." *International Journal of Food and Nutrition Research,*2:15. doi: 10.28933/ijfnr-2018-09-0301.

Talari, Aruna, and Devindra Shakappa. 2018. "Role of pigeon pea (*Cajanus cajan* L.) in human nutrition and health: A review." *Asian*

*Journal of Dairy and Food Research* 37: 212-20.doi:http://dx.doi.org/10.18805/ajdfr.DR-1379.

Tan, Ren Xiang, and Wen Xin Zou. 2001."Endophytes: a rich source of functional metabolites." *Natural product reports* 18: 448-59.doi:https://doi.org/10.1039/B100918O.

Tiwari, A. K., and A. K. Shivhare. 2016. "*Pulses in India: Retrospect and prospects*." Published by Director, Govt. of India, Ministry of Agri. & Farmers Welfare (DAC&FW), Directorate of Pulses Development, Vindhyachal Bhavan, Bhopal, MP.

Tiwari, Ashok Kumar, Bacha Abhinay, Katragadda Suresh Babu, Domati Anand Kumar, Amtul Zehra, and Kuncha Madhusudana. 2013."Pigeon pea seed husks as potent natural resource of anti-oxidant and anti-hyperglycaemic activity." *International Journal of Green Pharmacy (IJGP)* 7:252-57. Doi:10.4103/0973-8258.120247.

Tosh, Susan M., and Sylvia Yada. 2010."Dietary fibres in pulse seeds and fractions: Characterization, functional attributes, and applications." *Food Research International* 43: 450-60. doi:https://doi.org/10.1016/j.foodres.2009.09.005.

Uchegbu, Nneka N., and Charles N. Ishiwu. 2016. "Germinated Pigeon Pea (*Cajanus cajan*): a novel diet for lowering oxidative stress and hyperglycemia." *Food science & nutrition* 4: 772-77. doi: https://doi.org/10.1002/fsn3.343.

Venkidasamy, Baskar, Dhivya Selvaraj, Arti Shivraj Nile, Sathishkumar Ramalingam, Guoyin Kai, and Shivraj Hariram Nile.2019. "Indian pulses: A review on nutritional, functional and biochemical properties with future perspectives." *Trends in Food Science & Technology* 88: 228-42.doi:https://doi.org/10.1016/j.tifs.2019.03.012.

Vo, Thuy-Lan Thi, Nae-Cherng Yang, Shu-Er Yang, Chien-Lin Chen, Chi-Hao Wu, and Tuzz-Ying Song. 2020."Effects of *Cajanus cajan* (L.) millsp. roots extracts on the antioxidant and anti-inflammatory activities." *Chinese Journal of Physiology* 63:137-48.doi:https://doi.org/10.4103/cjp.cjp_88_19.

Wei, Zuo-Fu, Shuang Jin, Meng Luo, You-Zhi Pan, Ting-Ting Li, Xiao-Lin Qi, Thomas Efferth, Yu-Jie Fu, and Yuan-Gang Zu. 2013.

"Variation in contents of main active components and antioxidant activity in leaves of different pigeon pea cultivars during growth." *Journal of agricultural and food chemistry* 61: 10002-09.doi: https://doi.org/10.1021/jf402455m.

Wei, Zuofu, Yuangang Zu, Yujie Fu, Wei Wang, Meng Luo, Chunjian Zhao, and Youzhi Pan. 2013."Ionic liquids-based microwave-assisted extraction of active components from pigeon pea leaves for quantitative analysis." *Separation and Purification Technology* 102:75-81.doi:https://doi.org/10.1016/j.seppur.2012.09.031.

Wu, Nan, Kuang Fu, Yu-Jie Fu, Yuan-Gang Zu, Fang-Rong Chang, Yung-Husan Chen, Xiao-Lei Liu, Yu Kong, Wei Liu, and Cheng-Bo Gu. 2009."Antioxidant activities of extracts and main components of pigeonpea [*Cajanus cajan* (L.) Millsp.] leaves." *Molecules* 14:1032-43.doi:https://doi.org/10.3390/molecules14031032.

Wu, Nan, Yu Kong, Yujie Fu, Yuangang Zu, Zhiwei Yang, Mei Yang, Xiao Peng, and Thomas Efferth. 2011."In vitro antioxidant properties, DNA damage protective activity, and xanthine oxidase inhibitory effect of cajaninstilbene acid, a stilbene compound derived from pigeon pea [*Cajanus cajan* (L.) Millsp.] leaves." *Journal of agricultural and food chemistry* 59: 437-43.doi:https://doi.org/10.1021/jf103970b.

Zhang, Dong-Yang, Su Zhang, Yuan-Gang Zu, Yu-Jie Fu, Yu Kong, Yuan Gao, Jin-Tong Zhao, and Thomas Efferth. 2010."Negative pressure cavitation extraction and antioxidant activity of genistein and genistin from the roots of pigeon pea [*Cajanus cajan* (L.) Millsp.]." *Separation and Purification Technology* 74: 261-70.doi:https://doi.org/10.1016/j.seppur.2010.06.015.

Zhang, Meng-Di, Xue Tao, Rui-Le Pan, Li-Sha Wang, Chen-Chen Li, Yun-Feng Zhou, Yong-Hong Liao, Shan-Guang Chen, Qi Chang, and Xin-Min Liu. 2020."Antidepressant-like effects of cajaninstilbene acid and its related mechanisms in mice." *Fitoterapia* 141: 104450.doi:https://doi.org/10.1016/j.fitote.2019.104450.

Zhao, J., C. Li, W. Wang, C. Zhao, M. Luo, F. Mu, Y. Fu, Y. Zu, and M. Yao. 2013. "Hypocrealixii, novel endophytic fungi producing anticancer agent cajanol, isolated from pigeon pea (*Cajanus cajan* [L.] Millsp.).” *Journal of applied microbiology* 115:102-13.doi: https://doi.org/10.1111/jam.12195.

Zhao, JinTong, YuJie Fu, Meng Luo, YuanGang Zu, Wei Wang, Chun Jian Zhao, and Cheng Bo Gu. 2012.”Endophytic fungi from pigeon pea [*Cajanus cajan* (L.) Millsp.] produce antioxidant cajan instilbene acid.” *Journal of agricultural and food chemistry* 60:4314-19.doi:https://doi. org/10.1021/jf205097y.

Zu, Yuan-gang, Xiao-lei Liu, Yu-jie Fu, Nan Wu, Yu Kong, and Michael Wink. 2010. “Chemical composition of the SFE-CO2 extracts from *Cajanus cajan* (L.) Huth and their antimicrobial activity in vitro and in vivo.” *Phytomedicine* 17:1095-1101.doi:https://doi.org/10.1016/j. phymed.2010.04.005.

Zu, Yuan-gang, Yu-jie Fu, Wei Liu, Chun-lian Hou, and Yu Kong. 2006.”Simultaneous determination of four flavonoids in Pigeon pea [*Cajanus cajan* (L.) Millsp.] leaves using RP-LC-DAD.” *Chromatographia* 63: 499-505.doi:https://doi.org/10.1365/s10337-006-0784-z.

In: *Cajanus cajan*
Editor: Donald S. Wilkes
ISBN: 978-1-53619-134-9


***Chapter 3***

# IMPORTANCE AND USE OF PIGEON PEA (*CAJANUS CAJAN L*) IN FUNCTIONAL FOODS DESIGN

***Y. Barboza*[1] *and L. M. Medina*[2,*]**
[1]Laboratorio de Investigación y Desarrollo en Nutrición, Facultad de Medicina, Escuela de Nutrición y Dietética, Universidad del Zulia, Apartado, Maracaibo, Venezuela
[2]Departamento de Bromatología y Tecnología de los Alimentos, Edificio Darwin, Anexo, Campus Universitario Rabanales, Universidad de Córdoba, Córdoba, Spain

## ABSTRACT

Pigeon pea (*Cajanus cajan* (L.), among legumes, has an important role in the diet of many people in the world. It is one of the oldest food crops. Is the sixth important legume crop, with more than 7 million hectares planted area and ~4.89 million tones productions. Pigeon pea is rich in protein, carbohydrates, and dietary fiber, and a rich source of other bioactive components. Pigeon pea is a good source of dietary fiber and is low in fat, which helps in the maintenance of body weight and reduces the risk of cardiovascular diseases. Moreover, there are various potential health benefits due to the phytochemicals, such as phenolics, flavonoids, phytates, lectins, tannins, saponins, oxalates, enzyme inhibitors and phytosterols. In addition, pigeon pea is also rich in minerals (iron, sulphur, calcium, potassium, and manganese), water-soluble vitamins (thiamine, riboflavin, niacin) and some indispensable amino acids, such as leucine, lysine, phenylalanine, isoleucine, and valine. On the other hand, pigeon pea has been used in the formulation and evaluation of various products such as creams to spread with *Lactobacillus reuteri,* pure with fruits and biscuits among others. It has also been used to replace wheat flour in various products. In this review, we discuss the chemical composition, nutritional value, phytochemical components, health benefits and its usefulness in formulating functional foods.

**Keywords**: pigeon pea, germination; chemical composition; functional properties

## INTRODUCTION

The Fabaceae or Leguminosae (commonly known as the legume, pea, or bean) family is the third largest family of flowering plants, consisting of over 20,000 species. After cereals, pulses are one of the most economically important crop families in the world, representing about 15% of world's arable land (270–300 million hectares). Legumes are a nutritious staple of diets around the world. They are an inexpensive source of protein, vitamins, complex carbohydrates, fiber and several phytochemicals (Caprioli, et al., 2016; FAO, 2016).

To highlight their important contribution to the worldwide nutrition, low environmental impact of production and their adaptability to marginal soils and climates, the FAO declared 2016 as the International Year of Pulses (FAO, 2016). Pulses for human consumption cover crops as peas (*Pisum sativum*), beans (*Phaseolus* spp.; *Vicia faba*), chickpea (*Cicer arietinum*), pigeon pea (*Cajanus cajan*), lentils (*Lens esculenta*, *Ervum lens*), and lupin (*Lupinus* spp.) among others legume seeds widely produced as soybeans (*Glycine max*) (FAO, 1994; Giusti, et al., 2017).

Discrimination between legumes and pulses was determined by the Codex Alimentarius Commission of the FAO/WHO Food Standard, based on fat content (FAO/WHO, 2009). Legumes encompass oil seeds, including soybean and peanuts, while pulses are the dry seeds of leguminous plants. Cowpeas, pigeon peas, chickpeas, dry beans, dry broad beans, lentils, vetches, Bambara beans, lupins, and other minor pulses are the primary pulses grown worldwide.

Legumes have been a staple food included in the diet of a diversity of cultures around the world. Their high nutritional value and low-cost food made them an interesting source of bioactive compounds such as bioactive components and phytochemicals (Sánchez-Villegas, et al., 2018). Moreover, there are various potential health benefits due to the phytochemicals, such as phenolics, flavonoids, phytates, lectins, tannins, saponins, oxalates, enzyme inhibitors, phytosterols, and antimicrobial peptides, present in pulses. These phytochemicals have anti-inflammatory, anticancer, anti-microbial, and anti-ulcerative effects. In addition, pulses are also rich in vitamins [folate, thiamine (B1), riboflavin (B2), and niacin (B3)] and minerals (potassium, calcium, magnesium, phosphorus, and iron). Due to their high protein, content and low cost, pulses are referred to as 'poor people's meat (Venkidasamya, et al., 2019).

In fact, legumes are a rich source of essential nutrients and phytochemicals including natural antioxidants, which can provide beneficial effects on human health. The nutritional value of legumes is limited by the presence of antinutritional factors, such as, protease inhibitors, α-amylase inhibitors, phytic acid, lectins and saponins (Muzquiz et al., 2012). These compounds have been considered antinutrients because most of them can decrease the digestibility of proteins and carbohydrates and the bioavailability of vitamins and minerals.

For instance, phytic acid is considered an antinutrient compound due to its capacity to binds to mineral, protein and starch, thereby decreasing their bioavailability (Parca et al., 2018). However, some antinutrients may exert beneficial health effects at low concentrations. For instance, it has been reported that phytic acid has antioxidant DNA protective effects (Bhagyawant, et al., 2018). Furthermore, antinutrients in legumes are reduced using different methods including cooking, autoclaving, soaking, enzyme processing, fermentation and germination (Dida et al., 2018; Olika et al., 2019).

Legumes have been a staple food included in the diet of a diversity of cultures around the world. Their high nutritional value and low-cost food made them an interesting source of bioactive compounds such as bioactive components and phytochemicals (Sánchez et al., 2018). Many of these phytochemicals such as phenolic acids, anthocyanins, proanthocyanidins and flavonols, have been identified in different legumes (Kan et al., 2018).

There is growing evidence that high legume consumption is associated with a lower risk of suffer CVD and overall it is suggested that the inclusion of 150 g/day of cooked legumes in the diet is associated with lower mortality in the population. In Latin America countries, high legume intake equivalent to >86 g/day was associated with a 38% lower risk of myocardial infarction (Miller et al., 2017).The global increase in the prevalence of obesity and associated co-morbidities, such as type-2 diabetes (T2D), cardiovascular disease and cancer represents a serious public health problem. Multiple factors potentially contribute to the progression of cancer in obesity and T2D including hyperinsulinemia, hyperglycemia, dyslipidemia, inflammation and alterations in the composition of the gut microbiome (Sánchez et al., 2018).

In this context, as indicated above, legume consumption has been related to a decreased risk of 10 major chronic diseases including coronary heart disease, stroke, heart failure, diabetes, breast cancer, colorectal cancer, lung cancer, chronic obstructive pulmonary disease, dementia and depression (Kromhout et al., 2016). Various meta-analysis have also concluded that legume consumption reduces the risk of suffering colorectal cancer (Aune, et al., 2011; Wang, et al., 2013; Zhu et al., 2015).

A number of epidemiological studies have highlighted the inverse relationship between regular pulse consumption and risk of developing colorectal cancer, or even liver cancer (Zhang et al., 2013) or glioma (Benisi-Kohansal et al., 2016), although other studies did not find any effect of legume consumption on colorectal cancer incidence. Interestingly, although colorectal cancer incidence data have increased worldwide, inverse tendencies between incidence of colorectal cancer and pulses consumption can be clearly observed over the years (Vieira et al., 2017; World Cancer Research Fund/American Institute for Cancer Research, 2007).

Thus, in countries where pulse consumption was maintained or has increased, such as for example USA and Canada, colorectal cancer cases experienced a smaller increase as compared with those geographical areas where the decrease in the consumption of pulses was more pronounced (Zander et al., 2016). However, due to the low legume intake, the development of legume-based food products or functional ingredients could bring positive health benefits to consumers.

## *CAJANUS CAJAN*

Pigeon pea (*Cajanus cajan* -L.- Millsp.) among legumes, has an important role in the diet of many people in the world. It is one of the oldest food crops is the sixth important legume crop, with more than 7 million hectares planted area and ~4.89 million tones productions. India alone contributes over 90% of the world pigeon pea production. It is also a food crop in many other tropical countries and is commercially important in East Africa, the Caribbean and Latin America. It has low concentrations of fat, moderate amount of fiber, good amount of proteins and starch and a reasonably balanced range of all dietary essential minerals. (Olawuni, et al., 2012). Pigeon pea *Cajanus cajan* (L.) is the oldest crop in India and is commonly referred to as red gram or pigeon pea. The *C. cajan* cultivation area covers 11.8% of the total pulse growing area in India. *C. cajan* is grown both as a food crop and cover/forage crop (Tiwari et al., 2016).

Pigeon pea is a drought tolerant legume is grown mainly in the semi-arid tropics though it is well adapted to several environments (Troedson et al., 1990.), and can be grown in a wide range of soil textures, from sandy soils to heavy clays. It grows best at a soil pH of 5.0–7.0 but tolerates a wider range (4.5–8.4). It does well in low fertility soils, making it a favorite among subsistence farmers. As with most legumes, it does not tolerate waterlogged or flooded conditions for very long. Pigeon pea is very heat-tolerant and grows well in hot, humid climates; it thrives under annual rainfall between 24 and 40 inches (600–1000 mm). It is generally grown where the temperatures are in the range of 64–85$^0$F (18–30$^0$C), but under moist soil conditions it can withstand temperatures of 95$^0$F (35$^0$C) or more (Ravi et al., 2011).

Once established, it is one of the most drought tolerant of the legumes, and it can be grown in rained conditions or with minimal irrigation. In Hawaii, pigeon peas grow year-round at elevations ranging from sea level to 3000 ft, according to the USDA Natural Resources Conservation Service (NRCS) (Valenzuela et al., 2002). Pigeon pea remains one of the most drought-tolerant legumes (Valenzuela et al., 2002) and is often the only crop that gives some grain yield during dry spells when other legumes such as field beans will have wilted and perhaps dried up. The ability of pigeon pea to withstand severe drought better than many legumes is attributed to its deep roots (Flower et al., 1987) and osmotic adjustment (OA) in the leaves (Subbarao et al., 2000).

The legume also maintains photosynthetic function during stress better compared to other drought-tolerant legumes such as cowpea (*Vigna unguiculata* L. Walp.) (Lopez et al., 1987). Its unique polycarpic flowering habit further enables the crop to shed reproductive structures in response to stress (Mligo et al., 2005). Pigeon pea, has the ability to fix up to 235 Kg Nitrogen (N)/ha (Peoples et al., 1995) and produces more N per unit area from plant biomass than many other legumes. Nitrogen is one of the most abundant elements on earth (Vance, 2001) yet the most limiting nutrient for increasing crop productivity (Graham et al., 2003).

The N-fixing ability of pigeon pea is desirable for environmentally sustainable agricultural production (Peoples et al., 1995). While most legumes require inoculation to optimize their N-fixing ability, pigeon pea rarely needs inoculation because it can nodulate on *Rhizobium* that is naturally present in most soils (Faris, 1983). Pigeon pea offers the benefits of improving long-term soil quality and fertility when used as green manure (Onim et al., 1990). The legume also has the ability to reduce the level of root-knot nematodes in the succeeding crop when used as green manure (Daniel and Ong, 1990). Pigeon pea has been used successfully under coffee plantations as a cover crop to improve soil properties, reduce weed competition as well as act as a food source for predators (Venzon et al., 2006).

## Chemical Composition

The proximate composition of pigeon pea is 20–22% protein, 1.2% fat, 65% carbohydrate and 3.8% ash, as reported by FAO (1982) and Amarteifio et al., (2002). Is a valuable source of crude fiber, minerals (iron, sulphur, calcium, potassium, and manganese), water-soluble vitamins (thiamine, riboflavin, niacin) and some indispensable amino acids, such as leucine, lysine, phenylalanine, isoleucine, and valine (Sharma et al., 2011).

## Protein

Based on their solubility, seed proteins are classified into albumins (water extractable), globulins (soluble in weak salt solutions), prolamins (extractable in aqueous alcohol), and glutelins (soluble in weakly acidic or alkaline or dilute SDS) (Mandal, 2000). Globulins and albumins are the major protein components found in pulses. The major globulins found in pulses are legumin (11S) and vicilin (7S). Protease inhibitors, lectins, amylase inhibitors, and enzymatic proteins belong to the group albumins, which are also prevalent in pulses (Boye et al., 2010).

Plant proteins lack certain amino acids that are present in animal proteins, suggesting that plant proteins are inferior to animal proteins. Pulses contain other essential amino acids such as histidine, leucine, isoleucine, valine, threonine, phenylalanine, tryptophan, and lysine, but do not have sulphur-containing amino acids such as cysteine and methionine. Moreover, the types and concentrations of amino acids vary between pulses (Carbonaro, et al., 2012).

Albumins (rich in lysine and sulphur-containing amino acids) and globulins (aspartic acid and glutamic acid) have different amino acid profiles. However, pulse intake, combined with cereal intake, balances the amino acid deficiency (Vega et al., 2010). Some of the proteins found in pulses are considered biologically active peptides that exert various health-promoting activity such as angiotensin-converting enzyme (ACE) inhibition, immunomodulation, and anti-carcinogenic effects (Carbonaro et al., 2012).

Pigeon pea protein is rich in lysine, which is the first-limiting indispensable amino acid for humans in cereals such as wheat. Consequently, pigeon pea is usually combined with wheat to obtain a well-balanced protein diet. In addition to population growth and increased demand for affordable and accessible protein-rich foods for nutritional purposes, pigeon pea has been traditionally utilized as a source of dietary protein in many developing countries (Torres et al., 2006).

## Fibers

Like all legumes, pigeon pea is a good source of dietary fibers. Dietary fibers include resistant starch, nonstarch polysaccharides (cellulose, hemicellulose, pectin, gums, and b-glucans), nondigestible oligosaccharides, and lignin (Tharanathan et al., 2003). The ratio of soluble to insoluble fibers in legumes is comparable to that of grains (approximately 1:3 for both). High consumption of soluble fibers is associated with a decrease in serum total cholesterol, in LDL cholesterol, and is inversely correlated with coronary heart disease mortality rates (Buil-Cosiales et al., 2014, Brown et al., 1999). In addition, an important consumption of dietary fibers, in particular resistant starch, is related with improved glucose tolerance and insulin sensitivity as we have already mentioned throughout the chapter (Jenkins et al., 1980). It has been suggested that a state of satiety may be reached faster and last longer after ingestion of high-fiber foods, because they are bulkier and take longer to eat than lower fiber foods and delay gastric emptying.

Legumes contain a considerable amount of resistant starch, which is any starch that resists to digestion by amylase in the small intestine and progresses to the large intestine for fermentation by the gut bacteria (Yadav et al., 2010, Thorne et al., 1983). The resistant starch content of beans is much higher than in commonly consumed grains, most likely because of their high ratio of amylose to amylopectin; amylose is a no branched, linear polymer of glucose units that is less readily digested than amylopectin. The resistant starch content from legumes varies between 1.7 g/100 g of boiled weight in black beans and 4 g in navy beans and chickpeas and lentils with 2.6 and 3.4 g/100 g of boiled legumes (Yadav et al., 2010, Thorne et al., 1983).

In addition to a high resistant starch content, legumes also have a higher ratio of slow digestible to rapid-digestible starch, compared to other carbohydrate foods (Thorne et al., 1983). Legumes generally have a low glycemic index compared with other carbohydrate-rich foods, likely a result of both their resistant starch and fiber content (Jenkins et al., 1980). The glycemic index of legumes ranges from 29 to 38 compared with 50 for brown rice and 55 for rolled oats (Jenkins et al., 1980). The low glycemic index of legumes can potentially produce clinically relevant benefits related to CVD. Resistant starch is associated with reduced glycemic response, which can be beneficial to insulin-resistant individuals and those with diabetes (Park et al., 2004).

## Medicinal Applications

Considerable progress has been achieved regarding understanding pigeon pea or red gram's biological activity and medicinal applications. Flavonoids and stilbenes are found to be higher than in the leaves of *Cajanus cajan* compared to pods. In addition, tannins, saponins, resins, reducing sugars, and terpenoids are also present. (Kong et al., 2010). The presence of 2′-2′-methyl cajanone, 2′-hydroxy genistein, isoflavones, cajanin, cahanones, etc., results in antioxidant properties. Genistein and genistin are found in roots.

They also contain hexadecanoic acid, α-amyrin, β-sitosterol, pinostrobin, longistylin A, and longistylin C, which show antibacterial activity. *C. cajan* contains a significant amount of cajaninstilbene acid, pinostrobin, vitexin, and orientin, which are responsible for antiplasmodic activity (Dilipkumar et al., 2011). Pigeon pea is a rich source of secondary metabolites, which not only meet the needs of people's diet, but also enhance immunity, exhibit anti-inflammatory, anti-analgesic properties, and even play an important role in curing tumor (Zhang et al., 2013).

## Treatment to Increased Protein Digestibility of Pigeon Pea (*Cajanus cajan*)

Like other legumes, pigeon pea seeds contain antinutritional factors, such as phytic acid, polyphenols, saponins, protease inhibitors, phytolectins and oligosaccharides (Singh, 1988; Torres et al., 2006). These compounds can limit the utilization of pigeon pea seeds for human nutrition by impeding nutrient digestibility. For example, polyphenols and protease inhibitors have been demonstrated to decrease legume protein digestibility in humans, possibly due to their effect in decreasing protein bioaccessibility or inhibiting digestive proteases (Bressani et al., 1988). Within pigeon pea cultivars, anti-nutritional factors are mainly found among dark seeded genotypes that are typically grown in Asia. The native African pigeon pea types are largely cream or white seeded with relatively less antinutritional factors (Faris et al., 1990).

Some processing techniques, such as extrusion, soaking, cooking, germination, fermentation, sprouting, and microwave, can increase legume protein digestibility (Deng, et al., 2015; Nosworthy et al., 2018; Rathod et al., 2016; Singh, 1993; Sharma, et al., 2019a; Xu et al., 2019; Venkidasamy et al., 2019). In this sense, recent study indicates that the microwave as a promising technique for improving pigeon pea flour protein digestibility, with a potential to increase pigeon pea utilization in food product development.

In this work, the effects of physical treatments, including soaking, grinding, ultrasound and microwave, on the in vitro protein digestibility of pigeon pea flour were investigated. The results showed that the hypothesis only held true for microwave treatment, which significantly increased in vitro protein digestibility, despite its effect in decreasing the protein water solubility. This finding may be due to the changed starch structures in the flour or decreased β-sheet structure and increased random coil conformation of the proteins. Furthermore, the smaller particle size of the proteins resulting from the microwave treatment may have contributed to increasing the protein-protease interactions, thus improving protein digestibility (Xiaohong et al., 2020).

## Germination

Legume seeds are rich sources of dietary proteins for meeting human protein needs. However, the bioaccessibility of these proteins is hampered by the interaction of the proteins with other components of the seed matrix. Thus, improvement of protein digestibility is dependent on the hydrolysis of the indigestible proteins, deactivation of protease inhibitors, and improvement of protein solubility. Germination, a natural phenomenon in seed regeneration improves the total protein content of seeds when compared to non-germinated seeds. Furthermore, germination removes or represses protease inhibitors, thus improving protein digestibility and bioaccessibility (Ikenna et al., 2020).

Seed germination leads to regrowth characterized by the physiological appearance of sprouts, which is preceded by biochemical processes that activate the hydrolysis of stored nutrients. Germination is a complex process that occurs in three phases based on the seed water intake capabilities and dependent on the seed microstructure. The first phase is the imbibition stage, characterized by rapid water uptake; at the second phase, water uptake is reduced; and in the third phase, water uptake increases once again with the concomitant protrusion of the radicle from the seed coat (Bewley et al., 2013). Imbibition is necessary for the succeeding steps, as it enhances the activation of biochemical processes occurring in the second phase (suggested to be the most important phase of seed germination) and radicle emergence in the third phase (Wolny et al., 2018).

Germination was reported to repress the amount of phytate, tannins and trypsin inhibitors in different legume seeds (Sangronis et al., 2007; Chilomer et al., 2010) In the case of phytate, germination was described to reduce their seed content by activating the enzyme phytase and other phosphatases. This activation degrades phytate and yields important nutrients such as phosphate, inositol and other micronutrients (Bohn *et al.*, 2008). The phytate reductions were more pronounced (~60%) in peanut, chickpea and pigeon pea (*Cajanus cajan*) when compared to the other legumes evaluated.

## Bioactive Components

The total phenolic content (TPC) of germination pigeon pea during various germination conditions ranged from 6.32 to 8.36 mg GAE/g. Increase in germination time from 12 to 48 hr. and temperature from 25 to 35°C, significantly ($p < 0.05$) increased the TPC of pigeon pea. After 12 hr. of germination, the mean TPC increased by 11.73% which after 24,36, and 48 hr. of germination was increased by 20.11%, 23.81%, and 30.37%, respectively, as compared to control. Germination of pigeon pea at 25°C irrespective of time showed 13.09% higher TPC as compared to control. TPC further increased by 18.19% and 32.99%, when germination temperature was increased to 30 and 35°C, respectively. The interaction between germination time and temperature also had a significant ($p < 0.05$) effect on the TPC of pigeon pea. It was clear that increase in both time and temperature during germination had significantly ($p < 0.05$) increased the TPC of the pigeon pea (Uchegb et al., 2016; Sharma et al., 2019).

Shama et al., (2019) reported that increase in germination time and temperature modifies the bioactive components and nutritional digestibility of the pigeon pea. Studies have shown that increase in germination time from 12 to 48 hr. and temperature from 25 to 35°C, results in significant increase in accumulation of total phenolic and flavonoid content because of cell wall degrading enzymes. Germination for prolonged time at higher temperature also significantly increases the antioxidant potential and reducing power of the germinated pigeon pea. Increased activity of hydrolytic enzymes alters the structure of starch and proteins and thus enhanced in vitro starch and protein digestibility and lowers down the hot paste viscosity of germinated pigeon pea.

Piñero et al., (2013) investigated the combined effect of imbibition and germination on the quality of the pigeon pea (*Cajanus cajan* -L- Millsp.) in addition, found that combination of the soaked and germinated was effective in reducing protein content and total polyphenols ($p<0.05$). A decrease of 32% of phytic acid content on the fourth day of germination and 12 at 24 hours of soaking was observed. Seeds germinated over four days with a soaking process of 12 and/or 24 h could be used as an effective method to reduce the total polyphenols on flour seeds. As separate methods soaking process for 24 hours and germination for four days decrease phytic acid content. Therefore, germination could be further explored as a robust bioprocessing tool for improving the quality of legume seed proteins for food, biopharmaceutical and nutraceutical purposes.

## Products Formulated with Pigeon Peas

New legume-based products have been recently developed by emerging companies committed to promote human health. New brands in the food market have caught consumer attention by including legume grains as major ingredients. The interest in legumes is attributed to their bioactive components content and functional properties, such as solubility and water-binding capacity, which play an important role in formulation and processing of legume based-products (Foschia, et al., 2017).

Principal legumes used in the formulation of legume-based foods include common beans, lentils, pigeon peas, chickpeas, peas, soybeans and mung beans. Commercially available legume-containing food products are chips, croutons, pasta, snacks and chocolate. Pigeon pea have been used in the formulation of cookies, creams, compotes paste and pate (Rangel et al., 2017; Barboza et al., 2012; Torres et al., 2007; Parra et al., 2013; Piñero et al., 2014).

Cookies are widely consumed throughout the world. In fact, they represent the largest category of snack foods in most parts of the world (Lorenz, 1983). Okpala et al., (2011) produced cookies from blends of germinated pigeon pea, fermented sorghum and cocoyam flours. Cookies containing high levels of pigeon pea flour, had acceptable sensory scores. Increase in pigeon pea flour resulted in increase in the Biological Value and Net Protein Utilization.

Tiwari et al. (2011), prepared biscuits by substituting wheat flour with dehulled pigeon pea (*Cajanus cajan L*) flour or pigeon pea byproduct flour. Composite flour blends were prepared by substituting wheat flour (WF) with either dehulled pigeon pea flour or pigeon pea byproduct flour at 95:5, 90:10, 85:15, 80:20 and 75:25 incorporation levels. Protein content of fortified biscuits increased by 1.3 and 1.4 times respectively compared to control along with a significant increase in fiber content. Results indicate that good quality biscuits with increased levels of protein and fiber can be prepared by substituting wheat flour without significantly affecting the sensory quality of biscuits. This study demonstrates the potential feasibility of incorporating pigeon pea milling byproducts in the manufacture of biscuits.

On the other hand, Rangel et al. (2017) showed that, it is technologically feasible the use of pigeon pea meal as a partial substitute for wheat flour and oats to make cookies. The percentage of protein and fiber was increased as higher the level of substitution of wheat flour by pigeon pea meal, finding higher percentage of fiber in formulations where pigeon pea flour with whole grain was prepared. Sensory evaluation showed that all evaluated attributes (color, taste, smell, texture and overall appearance), without discriminating according cookie formulation, were accepted by respondents school.

In addition, pigeon pea flour is an excellent component in the snack industry and has been recommended as an ingredient to increase the nutritional value of pasta without affecting its sensory properties. In this study, the germinated pigeon pea flour can be an excellent ingredient and has been recommended to increase the nutritional value of semolina pasta without affecting the sensory properties.

Germination of pigeon peas appears to be an effective process for enhancement of chemical and nutritional parameters in this legume, particularly the increase in vitamins B2, E and C and the reduction of antinutrients, such as a-galactosides, inositol phosphates and trypsin inhibitor activity. Germinated seeds, by these benefits, were incorporated as high-protein ingredients (up to 10%) in pasta making resulting in products with good acceptability and larger amounts of protein, total available sugars, dietary fiber, micronutrients, vitamins and PER than pasta made from 100% semolina (Torres et al., 2007).

Among the main products formulated with pigeon peas is a spreading cream prepared with oats and *Lactobacillus reuteri.* This work was the first to investigate the potential use of *L. reuteri* ATCC 55730 in pigeon pea and oat-based products, and shows that this type of substrates is suitable and can support a high level of viable cells during refrigerated storage for 28 days. The viability of the microorganism was always above the recommended levels of $10^6$–$10^7$ cfu/g. All the creams had a high degree of likeness for all attributes evaluated. The addition of *L. reuteri* did not affect colour, flavour, texture and chemical composition of the creams. In addition, the combination of oats and pigeon pea is very helpful, because the amino acids of both types of food are combined to form a complete protein and could be used as a means of resolving the problem of malnutrition associated with the consumption of cereal-based foods. Hence, the addition of *L. reuteri* to creams resulted in a product with great potential as a functional product with excellent sensory characteristics. (Barboza et al., 2012).

Parra et al. (2013), also, evaluated the use of pigeon pea (*Cajanus cajan*) as an appropriate substrate in the production of a legume-based fermented product with *Lactobacillus acidophilus* ATCC 314 or *Lactobacillus casei* ATCC 393 and then to ascertain the effects of the addition of ingredients such as powdered milk and banana or strawberry sauce. The products were analyzed for viable cell counts, pH, and sensory attributes during product manufacture and throughout the refrigerated storage period at 3, 7, 14, 21, and 28 days.

On the other hand, a study carried out by Piñero et al. (2014) evaluated the incorporation of germinated pigeon pea flour and fermented with *Lactobacillus plantarum* on the improvement of the nutritional and functional properties of a low fat pate, compared with a control and with a meat standard product. They found that the addition of pigeon pea flours germinated for 4 days and fermented with L. *plantarum* as a partial substitute of a meat product to elaborate pates, allowed obtaining nutritional products with an adequate level of proteins and good digestibility.

Moreover, with low fat content and an optimum quantity of fiber and total minerals. Also increasing its functional properties when reducing the antinutrient phytic acid and increasing the natural antioxidant levels (total phenols). The germinated and fermented pigeon pea flour can be a good alternative as a partial substitute for meat in the preparation of meat products. Besides, pigeon pea have received a great attention in gluten-free food formulations, there is need to enhance its bioactive components and improve its in vitro digestibility through bioprocessing of the grain for its efficient utilization in variety of novel food for all ages with potential health-promoting compounds (Sharma et al., 2019).

## References

Amarteifio J O, Munthali D C, Karikari S K, MorakeT K. The composition of pigeon peas (*Cajanus cajan* (L.) Millsp.) grown in Botswana. *Plant Foods for Human Nutrition* 2002; 57: 173–177.

Aune D, Chan D S, Lau R, Vieira R, Greenwood D C, Kampman E, Norat T. Dietary fiber, whole grains, and risk of colorectal cancer: Systematic review and dose-response meta-analysis of prospective studies. *British Medical Journal* 2011; *343* d6617.

Barboza, Y., Márquez, E., Parra, K., Piñero, M., Medina, L. Development of a potential functional food prepared with pigeon pea (*Cajanus cajan*), oats and *Lactobacillus reuteri* ATCC 55730. *International Journal of Food Sciences and Nutrition* 2012; 63 (7): 813–820.

Benisi-Kohansal S, Shayanfar M, Mohammad-Shirazi M, Tabibi H, Sharifi G, Saneei P, Esmaillzadeh A. Adherence to the Dietary Approaches to Stop Hypertension-style diet in relation to glioma: A case–control study. *British Journal of Nutrition* 2016; 115: 1108–1116.

Bhagyawant S, Bhadkaria A, Gupta N, Srivastava N. Impact of phytic acid on nutrient bioaccessibility and antioxidant properties of chickpea genotypes. *Journal of Food Biochemistry* 2018. https://doi.org/10.1111/jfbc.12678.

Bewley J D, Bradford K, Hilhorst H, Nonogaki H. *Seeds: Physiology of Development, Germination and Dormancy* (Third). New York: Springer 2013.

Bohn L, Meyer A S, Rasmussen S K. Phytate: impact on environment and human nutrition. A challenge for molecular breeding. *Journal of Zhejiang University Science B* 2008; 9(3): 165–191. https://doi.org/10.1631/jzus.B0710640

Boye J, Zare F, Pletch A. Pulse proteins: Processing, characterization, functional properties and applications in food and feed. *Food Research International* 2010; 43: 414–431. ttps://doi.org/10.1016/j.foodres.2009.09.003.

Brown L, Rosner B, Willett W, Sacks F M. Cholesterol-lowering effects of dietary fiber: a meta-analysis. *Am J Clin Nutr* 1999; 69: 30-42.

Carbonaro M., Maselli P, Nucara A. Relationship between digestibility and secondary structure of raw and thermally treated legume proteins: A fourier transform infrared (FT-IR) spectroscopic study. *Amino Acids* 2012; 43: 911–921. https://doi.org/10.1007/s00726-011-1151-4.

Caprioli G, Giusti F, Ballini R, Sagratini G, Vila-Donat P, Vittori S. Lipid nutritional value of legumes: Evaluation of different extraction methods and determination of fatty acid composition. *Food Chemistry* 2016; 192: 965–971.

Chilomer K, Zaleska K, Ciesiolka D, Gulewicz P, Frankiewicz A, Gulewcz K. Changes in the alkaloid, A-galactoside and protein fractions content during germination of different Lupin species. *Acta Societatis Botanicorum Poloniae* 2010; 79(1): 11–20.

Daniel JN, Ong CK, 1990. Perennial pigeon pea: A multipurpose species for agroforestry systems. *Agroforestry Systems* 1990; 10: 113–129.

Deng Y, Padilla-Zakour O, Zhao Y, Tao S. Influences of high hydrostatic pressure, microwave heating, and boiling on chemical compositions, antinutritional factors, fatty acids, in vitro protein digestibility, and microstructure of buckwheat. *Food and Bioprocess Technology* 2015; 8 (11): 2235–2245. https://doi.org/10.1007/s11947- 015-1578-9.

Dida Bulbula D, Urga K. Study on the effect of traditional processing methods on nutritional composition and antinutritional factors in chickpea (*Cicer arietinum*). *Cogent Food & Agriculture* 2018. https://doi.org/10.1080/23311932.2017.1422370

Dilipkumar P, Pragya M, Neetu S, Ashoke K. Biological activities and medicinal properties of *Cajanus cajan* (L) Millsp. *Journal of Advanced Pharmaceutical Technology & Research,* 2011; 2: 207–214. https://doi.org/10.4103/2231-4040.90874.

Faris D G. ICRISAT's research on pigeon pea, In: *Grain Legumes in Asia*. Pp. 17–20. 1983 ICRISAT, Patancheru, India.

Faris DG, Singh U. Pigeon pea: Nutrition and products. In: Nene, Y.L., Hall, S.D., Sheila, V.K. (Eds.). *The Pigeon pea.* Wallingford, UK. CAB International 1990.

FAO. Legumes in human nutrition. Food and Agriculture Organization of the United Nations. *Food and Nutrition Series* 1982; No. 20 Rome.

FAO/WHO. Symposium on nutrition security for India, issues and way forward- nutrition strategies. *Indian National Science Academy Rome* August 2009; pp.3-4.

FAO, Food and Agriculture Organization. *International year of pulses* 2016. http://www.fao.org/pulses-2016/about/en/. Accessed 04.12.19.).

FAO, Food and Agriculture Organization (1994). Definition and classification of commodities. *Cereals and cereal products.* http://www.fao.org/es/faodef/fdef01e.htm, pulses and derived products. http://www.fao.org/waicent/faoinfo/economic/faodef/fdef 04e.htm. Accessed 04.12.19).

Flower D J, Ludlow M. Variation among accessions of pigeon pea (*Cajanus cajan*) in osmotic adjustment and dehydration tolerance of leaves. *Field Crops Research* 1987; 17: 229–243.

Giusti F, Caprioli G, Ricciutelli M, Vittori S, Sagratini G. Determination of fourteen polyphenols in pulses by high performance liquid chromatography-diode array detection (HPLC-DAD) and correlation study with antioxidant activity and colour. *Food Chemistry* 2017; 221, 689–697.

Graham P H, Vance C P. Legumes: Importance and constraints to greater use. *Plant Physiology* 2003; 131: 872–877.

Foschia M, Horstmann S W, Arendt E K, Zannini E. Legumes as Functional Ingredients in Gluten-Free Bakery and Pasta Products. *Annual Review of Food Science and Technology* 2017. https://doi.org/10.1146/annurev-food-030216-030045.

Ikenna C. Ohanenye A, Chukwunonso E C, Ejike C, Udenigwe. Germination as a bioprocess for enhancing the quality and nutritional prospects of Legume proteins. *Trends in Food Science & Technology* 2020. https://doi.org/10.1016/j.tifs.2020.05.003.

Jamdar S N, Deshpande R, Marathe S A. Effect of processing conditions and in vitro protein digestion on bioactive potentials of commonly consumed legumes. *Food Bioscience* 2017; 20: 1–11. https://doi.org/10.1016/j.fbio.2017.07.007.

Jeong D, Han J A, Liu Q, Chung H J. Effect of processing, storage, and modification on in vitro starch digestion characteristics of food legumes: A review. *Food Hydrocolloids* 2019. https://doi.org/10.1016/j.foodhyd.2018.12.039.

Kan L, Nie S, Hu J, Wang S, Cui S, Li Y, Xie M. Nutrients, phytochemicals and antioxidant activities of 26 kidney bean cultivars. *Food and Chemical Toxicology* 2017; 108: 467– 477. https://doi.org/10.1016/j.fct.2016.09.007.

Kong Y, Fu Y J, Zu Y G, Chang F R, Chen YH, Liu XL. Cajanuslactone, a new coumarin with anti-bacterial activity from pigeon pea [*Cajanus cajan* (L.) Millsp.] Leaves. *Food Chemistry* 2010; 121: 1150–1155. https://doi.org/10.1016/j.foodchem.2010.01.062.

Kromhout D, Spaaij C J, Goede J, Weggemans R M, Brug J, Geleijnse J M, Zwietering M. The 2015 Dutch food-based dietary guidelines. *European Journal of Clinical Nutrition* 2016; 70: 869–878.

Lopez, F.B, Setter T.L, McDavid C.R. Carbon dioxide and light response of photosynthesis in cowpea and pigeon pea during water deficit and recovery. *Plant Physiology,* 1987; 85: 990–995.

Miller V, Mente A, Dehghan M, Rangarajan S, Zhang X, Swaminathan S, Mapanga R. Fruit, vegetable, and legume intake, and cardiovascular disease and deaths in 18 countries (PURE): a prospective cohort study. *The Lancet* 2017; 390: 2037–2049. https://doi.org/10.1016/S0140-6736(17)32253-5.

Mligo J K, Craufurd P Q. Adaptation and yield of pigeon pea in different environments in Tanzania. *Field Crops Research* 2005; 94: 43–53.

Muzquiz M, Varela A, Burbano C, Cuadrado C, Guillamón, E, Pedrosa M. Bioactive compounds in legumes: Pronutritive and antinutritive actions. Implications for nutrition and health. *Phytochemistry Reviews* 2012; https://doi.org/10.1007/s11101-012-9233-9.

Nosworthy M, Medina G, Franczyk A, Neufeld J, Appah P, Utioh A, House J. Effect of processing on the in vitro and in vivo protein quality of beans (*Phaseolus vulgaris* and *Vicia faba*). *Nutrients* 2018; 10(6): 671-676. https://doi.org/10.3390/ nu10060671.

Olawuni I, Ojukwu M, Eboh B. Comparative study on the physico-chemical properties of pigeon pea (*Cajanus cajan*) flour and protein isolate. *International Journal of Agricultural and Food Science* 2012; 2(4): 121–126.

Olika E, Abera S, Fikre A. Physicochemical properties and effect of processing methods on mineral composition and antinutritional factors of improved chickpea (*Cicer arietinum* L.) Varieties Grown in Ethiopia. *International Journal of Food Science* 2019. https://doi.org/10.1155/2019/9614570.

Onim J F, Mathuva M, Otieno K, Fitzhugh H A. Soil fertility changes and response of maize and beans to green manures of leucaena, sesbania and pigeon pea. *Agroforestry Systems* 1990; 12: 197– 215.

Parra K, Ferrer M, Piñero M, Barboza Y, Medina L. Use of *Lactobacillus acidophilus* and *Lactobacillus casei* for a Potential Probiotic Legume-Based Fermented Product Using Pigeon Pea (*Cajanus cajan*). *Journal of Food Protection* 2013; 76 (2): 265–271.

Parca F, Koca Y, Unay A. Nutritional and Antinutritional Factors of Some Pulses Seed and Their Effects on Human Health. *International Journal of Secondary Metabolite* 2018. https://doi.org/10.21448/ijsm.488651.

Park O J, Kang N E, Chang M J, Kim W K. Resistant starch supplementation influences blood lipid concentrations and glucose control in overweight subjects. *J Nutr Sci Vitaminol* 2004; 50: 93-99.

Peoples M B, Herridge D F, Ladha J K. Biological nitrogen fixation: An efficient source of Nitrogen for sustainable agricultural production? *Plant and Soil* 1995; 174: 3–28.

Piñero M P, Parra K, Barboza Y, Pérez M, Ortega J. Combined effect of imbibition and germination on the quality of the pigeon pea (*Cajanus cajan* (L) Millsp.). *Ciencia* 2013. 21(4): 192 - 200.

Piñero M P, Parra K, Barboza Y, Arévalo E, Méndez S. Potential use of germinated and fermented pigeon pea flour in the manufacture of a pate. *Rev. Fac. Agron.* (LUZ) 2014, Supl. 1: 678-687.

Prodanov M, Sierra I, Vidal-Valverde C. Influence of soaking and cooking on the thiamin, riboflavin and niacin contents of legumes. *Food Chemistry* 2004; 84(2): 271–277. https://doi.org/10.1016/s0308-8146 (03)00211-5.

Ravi R K, Krishna K, Naik G. Variation of sensitivity to drought stress in pigeon pea (*Cajanus cajan* [L.] Millsp) cultivars during seed germination and early seedling growth. *World Journal of Science and Technology* 2011; 1(1): 11-18.

Sánchez-Villegas A, Sánchez-Tainta A, Murphy K J, Marques-Lopes I, Sánchez-Tainta, A. Cereals and Legumes. *The Prevention of Cardiovascular Disease through the Mediterranean Diet* 2018; 111–132. https://doi.org/10.1016/B978-0-12-811259-5.00007-X.

Sangronis E Ã, Machado C J. Influence of germination on the nutritional quality of Phaseolus *vulgaris* and *Cajanus cajan. Swiss Society of Food Science and Technology* 2007; 40: 116–120. https://doi.org/10.1016/j.lwt.2005.08.003.

Sharma S, Agarwal N, Verma P. Pigeon pea (*Cajanus cajan L.*): A hidden treasure of regime nutrition. *Journal of Functional and Environmental Botany* 2011; 1(2): 91–101. https://doi.org/10.5958/j.2231-1742.1.2.010.

Sharma S, Singh A, Singh B. Characterization of in vitro antioxidant activity, bioactive components, and nutrient digestibility in pigeon pea (*Cajanus cajan*) as influenced by germination time and temperature. *Journal of Food Biochemistry* 2019a; 43(2): 12706. https://doi.org/10.1111/jfbc.12706.

Sharma S, Singh A, Singh B. Effect on germination time and temperature on techno-functional properties and protein solubility of pigeon pea (*Cajanus cajan*) flour. *Quality Assurance and Safety of Crops & Foods 2019b;* 11(3): 305–312. https://doi.org/ 10.3920/QAS2018.1357.

Singh U. Protein quality of pigeon pea (*Cajanus cajan L.*) Millsp.) as influenced by seed polyphenols and cooking process. *Plant Foods for Human Nutrition* 1993; 43(2): 171–179. https://doi.org/10.1007/bf01087921.

Subbarao G V, Chauhan Y S, Johansen C. Patterns of osmotic adjustment in pigeon pea — its importance as a mechanism of drought resistance. *European Journal of Agronomy* 2000; 12: 239–249.

Tharanathan R N, Mahadevamma S. Legumes—a boon to human nutrition. *Trends Food Sci Tech* 2003; 14: 507-18.

Thorne M J, Thompson L U, Jenkins D J. Factors affecting starch digestibility and the glycemic response with special reference to legumes. *Am J Clin Nutr* 1983; 38:481-488.

Tiwari B K, Brennan C S, Jaganmohan, R, Surabi A, Alagusundaram K. Utilization of pigeon pea (*Cajanus cajan L*) byproducts in biscuit manufacture. *LWT - Food Science and Technology* 2011; 44: 1533-1537. doi:10.1016/j.lwt.2011.01.018.

Torres A, Frías J, Granito M, Vidal-Valverde C. Fermented pigeon pea (*Cajanus cajan*) ingredients in pasta products. *Journal of Agricultural and Food Chemistry* 2006; 54(18): 6685–6691. https://doi.org/10.1021/jf0606095.

Tiwari A K, Shivhare A. K. *Pulses in India: retrospect and prospects. Published by Director, Govt. of India, Ministry of Agri. And Farmers Welfare (DACandFW)*. Publication No: DPD/Pub 2016; 1 Vol. 2/2016, 2, 1–21 https://farmer.gov.in/ imagedefault/prospects_2017.pdf.

Uchegbu N, Ishiwu C N. Germinated Pigeon Pea (*Cajanus cajan*): A novel diet for lowering oxidative stress and hyperglycemia Food. *Science and Nutrition* 2016; 4: 772–777. https://doi.org/10.1002/ fsn3.343.

Valenzuela H, Smith J. *Pigeon pea Sustainable Agriculture Green Manure Crops Aug.1-3- SA-GM-8* Published by the College of Tropical Agriculture and Human Resources (CTAHR): 2002; 1-3.

Vance C.P. Symbiotic nitrogen fixation and phosphorus acquisition: Plant nutrition in a world of declining renewable resources. *Plant Physiology* 2001; 127: 390–397.

Venkidasamy B, Selvaraj D, Nile A S, Ramalingam S, Kai G, Nile S H. Indian pulses: A review on nutritional, functional and biochemical properties with future perspectives. *Trends in Food Science & Technology* 2019; 88: 228–242. https://doi.org/10.1016/j.tifs.2019.03.012.

Vieira A R, Abar L, Chan D S, Vingeliene S, Polemiti E, Stevens C, Greenwood D, Norat T. (2017). Foods and beverages and colorectal cancer risk: A systematic review and meta-analysis of cohort studies, an update of the evidence of the WCRFAICR Continuous Update Project. *Annals of Oncology* 2017; *28:* 1788–1802.

Wang Y, Wang Z, Fu L, Chen Y, Fang J. Legume consumption and colorectal adenoma risk: A meta-analysis of observational studies. *PLoS ONE* 2013; 8(6): 67335.

Wolny E, Betekhtin A, Rojek M, Braszewska-zalewska A, Lusinska J, Hasterok R. Germination and the Early Stages of Seedling Development in *Brachypodium distachyon*. *International Journal of*

*Molecular Sciences* 2018; 19: 1–14. https://doi.org/10.3390/ijms 19102916

World Cancer Research Fund/American Institute for Cancer Research. *Food, Nutrition, Physical Activity, and the Prevention of Cancer: A Global Perspective* 2007; ISBN: 978-0-9722522-2-5.

Yadav B S, Sharma A, Yadav R B. Resistant starch content of conventionally boiled and pressure-cooked cereals, legumes and tubers. *J Food Sci Technol* 2010; 47: 84-88.

Zhang D Y, Zu Y, Fu Y J, Luo M, Wang W, Gu C B, Yao X H. 2013. Application of immobilized enzymes to accelerate the conversion of genistin to genistein in pigeon pea root extracts and the evaluation their antioxidant activity. *Industrial crops and products* 2013; 42: 409-415.

Zhang W, Xiang Y B, Li H, Yang G, Cai H, Ji B T, Shu X O. Vegetable-based dietary pattern and liver cancer risk: Results from the Shanghai Women's and Men's Health Studies. *Cancer Science* 2013; 104: 1353–1361.

Zhu B, Sun Y, Qi L, Zhong R, Miao X. Dietary legume consumption reduces risk of colorectal cancer: Evidence from a meta-analysis of cohort studies. *Scientific Reports – Nature* 2015; 5: 8797.

Xiaohong S, Ikenna C, Ohanenyea T, Ahmeda C, Udenigwea. Microwave treatment increased protein digestibility of pigeon pea (*Cajanus cajan*) flour: Elucidation of underlying mechanisms. *Food Chemistry* 2020; 329 127196. https://doi.org/10.1016/j.foodchem.

Xu M, Jin Z, Simsek S, Hall C, Rao J, Chen B. Effect of germination on the chemical composition, thermal, pasting, and moisture sorption properties of flours from chickpea, lentil, and yellow pea. *Food Chemistry* 2019; 295: 579–587. https://doi.org/10.1016/j.foodchem. 2019.05.167.

In: *Cajanus cajan*
Editor: Donald S. Wilkes
ISBN: 978-1-53619-134-9


*Chapter 4*

# *CAJANUS CAJAN* (L.) MILLSP. CULTIVATION, USE AND NUTRITION

***P. T. Patel[1,*], PhD and A. B. Khatri[2]***
[1]Associate Research Scientist Department of Seed Technology, Sardarkrushinagar Dantiwada Agricultural University, Sardarkrushinagar, Gujarat, India
[2]Department of Genetics and Plant Breeding, C. P. College of Agriculture, Sardarkrushinagar Dantiwada Agricultural University, Sardarkrushinagar, Gujarat, India

## ABSTRACT

Pigeon pea [*Cajanus cajan* (L.) Millsp.] is a versatile pulse crop with high nutritive content. It can be grown in tropical and sub-tropical areas of the world. It can be cultivated in black cotton soils and well drained with a pH ranging from 7.0-8.5. As well as being a pulse crop pigeon pea fixes atmospheric nitrogen into soil and increases the fertility. India is major producer in the world and accounts for more than 70 percent of the production. Pigeon pea production is slowly increasing due to enormous breeding efforts like exploitation of heterosis by development of hybrids

* Corresponding Author's E-mail: ptpatelsdau@gmail.com.

as well as male sterile lines and restorer lines in recent years. Pigeon pea is consumed as a raw immature seed, mature seed and processed dal. It is also used for animal feed and host material for lac production. Pigeon pea has also been exploited as a medicinal plant as the leaf extract contains an anti-microbial compound. It is an excellent source of protein (18-25%), starch (45-60%), crude fiber (8.2%) and fat (2.3%). It contains the best amino acid balance to fulfil daily nutrient requirement as well as minerals like calcium and magnesium in good amounts. Immature pigeon pea seed contains a high amount of vitamin A and vitamin C. Pigeon pea has excellent potential for the improvement of production and nutrient arability especially in poor developing countries due to its natural behavior.

**Keywords**: pigeon pea, *Cajanus cajan*, male sterility, CMS, CGMS, hybrids, use, nutrition, cultivation

## 1. INTRODUCTION

Pigeon pea [*Cajanus cajan* (L.) Millsp.] is one of the major pulse crops of the tropics and sub tropics including India, Australia, Hawaii, Uganda, Italy, America, East and West Indies and South-East Africa. Pigeon pea is a short-lived perennial shrub (Van der Maesen et al. 1980), which plays a major role in improving soil nutrition by nourishing soil with atmospheric nitrogen fixation. It is known by various names, such as red gram and congo bean (English), tur and arhar (Hindi), guand (Portuguese), gandul (Spanish), poid d'Angole and poid de Congo (French) and ervilba de Congo in Angola, and is grown primarily as a food crop (Upadhyaya et al., 2013). It is the sixth most important grain legume in the world and second most important pulse crop after chickpea in India which is grown in the Indian subcontinent accounting for the 90% of the world's crop.

Pigeon pea is often cross-pollinated, with an insect-aided natural out-crossing range from 20% to 70% with chromosome number $2n=2x=22$ and genome size 1C = 858 Mbp. It belongs to the family *Leguminosae*, subfamily *Papilionoideae*, tribe *Phaseoleae* and the subtribe *Cajaninae*. The tribe *Phaseoleae* comprises many edible bean species (*Phaseolus*, *Vigna*, *Cajanus*, *Lablab*, etc.) of which the members of subtribe *Cajaninae* are well distinguished by the presence of vesicular glands on the leaves, calyx and pods. Currently, 11 genera are grouped under the subtribe *Cajaninae*, including *Rhynchosia* Lour., *Eriosema* (DC.), G. Don, *Dunbaria*, W. & A. and *Flemingia* Roxb. ex Aiton, but the cultivated pigeon pea *C. cajan* is the only domesticated species in *Cajaninae*. The word '*Cajanus*' is derived from a Malay word 'katschang' or 'katjang' meaning pod or bean. The revised genus *Cajanus* currently comprises 18 species from Asia, 15 species from Australia and 1 species from West Africa. Of these, 13 are found only in Australia, 8 in the Indian subcontinent, and 1 in West Africa, with the remaining 14 species occurring in more than 1 country. Based on growth habit, leaf shape, hairiness, structure of corolla, pod size and presence of strophiole, van der Maesen (1980) grouped the genus *Cajan* into six sections. (Upadhyaya et al., 2013).

Pigeon pea is is an erect perennial, warm-season crop that is widely grow shrub that can grow to 12 ft tall, but usually only reaches 3 to 6 ft. The ribbed stem grows upright and is covered in short, soft hairs (pubescent) and is woody at the base. Its deep tap root is fast-growing. The pubescent, stalked leaflets are 2 to 4 inches long (5–10 cm) and ¾ to 1½ inches wide (2–4 cm), with minute resinous glands underneath. The bi-laterally symmetrical, bell-shaped blooms are yellow or yellow and red, may be in pairs, and grow in the angle between stem and leaf on an unbranched inflorescence. The upper two lobes are united and two-toothed, and the lower lip is smaller and three-toothed. The lower bracts fall off early. The two-valve, pointed seedpods are produced in clusters, and mottled red. They are 2 to 3.5 inches (5–9 cm) long, ½ inch (12 mm) wide, flat, covered with soft hairs, and taper to a sharp point. The round or oval seeds may be light beige to dark brown. The 2 to 9-seeded pods do not shatter in the field. *C. cajan* has a deep taproot system that grows large, cylindrical nodules. This nodulation can be initiated by a variety of rhizobial strains (Sheahan, 2012).

In world pigeon pea is grown in the nearly 70 lakh ha with production of 59.6 lakh tonnes with 852 kg/ha productivity. In the world alone India contribute more than 70 *per cent* production of pigeon pea with 42.9 lakh tonnes in 55.8 lakh ha area with 768 kg/ha productivity. Myanmar, Malawi, United Republic of Tanzania, Haiti, Dominican Republic, Nepal and Uganda are also a good contributor of pigeon pea production (FAOSTAT, 2018). It is grown in almost all the states of India, but the major states in terms of area and production are Maharashtra, Uttar Pradesh, Madhya Pradesh, Karnataka, Gujarat, and Andhra Pradesh which together account for world's 87% of the area and 83.8% of the production.

Pigeon pea seeds are highly nutritious. The mature seeds contain 18.8% protein, 53% starch, 2.3% fat, 6.6% crude fiber and 250.3 mg per 100 g minerals. As a perennial shrub, pigeon pea has many advantages over annual legumes in that several harvests are possible and the capacity to contribute to enhance soil fertility is much higher. Pigeon pea has high tolerance to drought stresses, high biomass productivity, which is mainly used as fodder, and provides the most nutrient and moisture contributions to the soil with climate variability and the occurrence of prolonged drought, pigeon pea offers resilience to cropping systems and its cultivation is expected to expand to new areas. *C. cajan* can be grown as a forage intercropped with sorghum and/or millet. The deep taproot of *C. cajan* draws water from deeper soil depths than most legumes, so will not interfere with the water uptake of other crops and grasses. It is not generally seeded as forage with other legumes like cowpea (*Vigna unguiculata*), but mainly with grasses (Cook et al., 2005).

## 2. Cultivation

Pigeon pea is cultivated in tropical and sub-tropical area of world. The cultivation of pigeon pea requires a temperature ranging from 26°C to 30°C in the rainy season (June to October) and 17°C to 22°C in the post rainy (November to March) season. Pigeon pea is very sensitive to low radiation at pod development, therefore flowering during the monsoon and cloudy weather, leads to poor pod formation. It has been estimated that approximately 40 kg/ha of the residual nitrogen is left in the field by pigeon pea crop which resulted from the leaf fall and nitrogen fixation.

**Table 1. List of male sterile cytoplasm with source**

| No | Cytoplasm | Source | Reference |
|---|---|---|---|
| 1. | $A_1$ | *Cajanus sericeus* | Saxena, et al., 1997 |
| 2. | $A_2$ | *Cajanus scarabaeoides* | Tikka et al., 1997 |
| 3. | $A_3$ | *Cajanus volubilis* | Wanjari et al., 1999 |
| 4. | $A_4$ | *Cajanus cajanifolius* | Saxena et al., 2005 |
| 5. | $A_5$ | *Cajanus cajan* | Mallikarjuna and Saxena, 2005 |
| 6. | $A_6$ | *Cajanus lineatus* | Saxena et al., 2010 |
| 7. | $A_7$ | *Cajanus platycarpus* | Saxena et al., 2010 |

World production of pigeon pea continuously increased since 1960. It was nearby 20-21 lakh tonnes until 1984, then it increased to nearby 30 lakh tonnes annually. Sudden quantum jump in world production observed from 2006 which reach to nearly 40 lakh tonnes annually and in last recent years the pigeon pea production attained production near to 60 lakh tonnes which shows the progress of continuous breeding effort to sustain the production as well as quality. Similarly, the area grown under the pigeon pea increased from 27.2 lakh ha to 70 lakh ha nearly. The extensive breeding effort such as development of CGMS based male sterile system for hybrid seed production in pigeon pea leads to increase the productivity from 560 kg/ha in 1962 to 850 kg/ha 2018. (FAOSTAT, 2018).

Since the big challenge was to develop the efficient hybrid seed production method based on male sterility to exploit the heterosis present in pigeon pea germplasm. The first such GMS based hybrid from ICRISAT was released as ICPH-8 in 1991 which was able to yield more 30-35 *per cent* than many hybrids based on GMS released but the main constrain to adapt was costly seeds and hard to maintain purity inherently associated with any GMS system (Saxena, 2008). To overcome such constrains of GMS based hybrid system scientists focused to develop CGMS system based hybrid and efforts outcome with development of cytoplasmic male sterile lines having cytoplasm *viz.* $A_1$, $A_2$, $A_3$, $A_4$, $A_5$, $A_6$, $A_7$ ($A_1$ to $A_7$). Among this $A_2$ and $A_4$ cytoplasm believed to be stable as it stable over environment and having many restorer available (Tikka et al., 1997; Saxena et al., 2005; Patel and Tikka, 2014). The development of stable

CGMS based hybrid systems helped to provide quality hybrid seeds which enhanced the productivity of pigeon pea at farmer's level.

Pigeon pea is successfully grown in black cotton soils, well drained with a pH ranging from 7.0-8.5. It responds well to properly tilled and well drained seedbed. The seeding rate of pigeon pea depends on the desired plant density for a genotype (early, medium or late), cropping system (pure crop, mixed crop, or inter crop), germination rate of seed and mass of seed. Pigeon pea is commonly intercropped with a wide range of crops. In India, it was estimated that 80 - 90% of the pigeon pea were intercropped i.e.,

a) with cereals: sorghum, maize, pearl millet, finger millet and rain-fed rice
b) with legumes: groundnut, cowpea, mung bean, black gram, soybean and
c) with long-season annuals: caster, cotton, sugarcane, and cassava.

The important diseases of Pigeon pea are wilt, sterility mosaic disease, Phytophthora blight, Alternaria blight, Powdery mildew. While pod borer, Tur Pod fly, Plume Moth, Pod-sucking bugs are important pest of pigeon pea. Among them, sterility mosaic virus disease is one of the most destructive in India and Nepal as it can causing yield losses up to 95 *per cent* (Ganapathy et al., 2011; Reddy and Nene, 1981). The early stage (<45-days old plants) of infection results into 95 to 100 *per cent* yield losses as reported by Reddy et al. (1990) and Kulkarni et al. (2003). The disease results in 100 *per cent* yield loss when symptoms appear at the pre-flowering and podding stage, whereas, at maturity stage loss could be of 67 *per cent* and at pre-harvest stage up to 30 *per cent.* Seeds from partially infected plants appear discolored, shrivelled and results into 20 *per cent* reduction in dry weight while annual yield loss exceeds US $ 100 million in India alone (Kumar et al., 2000). Manifestation of PSMD chiefly depends on the availability of mite populations (Singh et al., 1999). The mite populations are usually positively correlated with rainfall, relative humidity and lower temperature. PPSMV is transmitted by an eriophyid mite (*Aceria cajani* Channabasavanna). The disease is characterized by the symptoms like bushy and pale green appearance of plants followed by reduction in leaf size, excess vegetative growth, increasing number of secondary branches and mosaic mottling of leaves and finally partial or complete cessation of reproductive structures. Sometimes, some parts of the plant may show disease symptoms and other parts may remain unaffected (Kumar et al., 2003). Currently wilt and PSMD resistant/tolerant varieties available which can be very useful to reduce yield loss such as BSMR 736, 853, 846, ICPL 96053, BDN 2010, ICPL 43, 44, IPA 203, 204, 234 and IPH 09-5 based on suitability of region.

## 3. USES

Pigeon pea is used in many ways somewhere it consumed raw green pods, somewhere it used as processed dal and the herb is used for the forage purpose as it is very nutritive to feed livestock. Pigeon pea seed used as whole seed. The dry pigeon pea seeds are soaked overnight and cooked with salt and spices. The boiled whole seeds are sometimes fried with spices and eaten with cereals, particularly in Africa. Foods such as *Bongko* and *Brubus* made with the whole seed are popular in Central Java. Other dishes and snacks are *rempeyek, serundeng,* and *gandasturi* (Damardjati and Widowati 1985). The seeds of pigeon pea are soaked in water and allowed to sprout. The sprouted seeds are eaten raw or cooked. The another way of consumption of pigeon pea is *Tempeh.* This is prepared in combination with soybean. Tempeh is prepared by fermenting soaked, denuded, and cooked pigeon pea seed with a *Rhizopus* mould. The seed is spread on a mesh, and the mould is allowed to grow on the surface and through the seed, forming a compact cake. The tempeh cake is cut into pieces and fried before eating. Ketchup (sauce) also can be made and used. Sometimes seeds of pigeon pea is used as canned seed or as roasted pigeon pea (especially in eastern India). The cotyledons of dry seeds excluding seed coat is called *dhal.* In India and many Asian countries, pigeon pea is mainly consumed as *dhal.* Dhal is popular because it takes less time to cook and has acceptable appearance, texture, palatability, digestibility and overall nutritional quality. Dal is cooked to make a thick soup primarily for mixing with rice. Similarly its used as the green seed as a vegetable as its more nutritious than the dried or processed pigeon pea. Immature pods are harvested before the seeds are distinct, and cooked like french beans in curries (Morton 1976). Such pods are also used as salads (Faris and Singh 1990).

### 3.1. Animal Feed

The dry leaves and the left over pods at threshing of the crop are used as feed for animals. The byproducts of seed coats, broken bits, and powder from the *dhal* mill collectively are called *'Chuni'*. It is a valuable food for milch cattle (Pathak 1970). Dry leaves of pigeon pea provide a good substitute for alfalfa in animal feed. In India it is mixed with wheat straw to feed cattle. Dry leaves were found to be a useful replacement for alfalfa as a source of carotene and other essential nutrients in chicken rations (Squibb et al. 1950). Being a perennial crop with large biomass production with a high level of nitrogen under low input conditions there is considerable focus on using pigeon pea as a fodder supplement, particularly in the areas where soybean does not grow well and soybean meal is imported for animal feed (Wallis et al. 1988). The long-duration genotypes were better adapted to cutting as long as lower leaves remained on the stubble.

### 3.2. Other Uses of Pigeon Pea

The wood (stem and branches) yield of pigeon pea in northern India is 6-10 t per ha$^{-}$from short-duration genotypes (ICRISAT 1984) and 3-6 t per ha from medium-duration genotypes in central and southern India (Jain et al. 1987). The thin branches are used in several ways. It is used as fuel for cooking. The straight branches are useful as light construction material for roofing, wattling on carts, tubular wicker-work lining for walls, and baskets (Pathak 1970). These are also useful for temporary fencing, hut construction, and binding material. Small baskets to carry the farm produce are made from the branches of pigeon pea after soaking these in water for 36 hours. Medium-sized, thin, and straight main shoots (after removing branches) are used to make big baskets. The big baskets are used as storage bins. The storage bins are plastered using a slurry of soil, dung, and benzene-hexa-chloride. Pigeon pea wood was found useful in production of pulp to make good quality writing and printing papers (Akhtaruzzaman et al. 1986).

### 3.3. Host for Producing Lac and Silk

Pigeon pea plants act as host for scaled insects producing lac in north Bengal and Assam in India, Thailand, Vietnam, and China (Faris and Singh 1990). Pigeon pea has the advantage of a short life cycle in comparison to other lac host plants. However, lac from pigeon pea does not bleach well, therefore it is considered inferior in quality (Macmillan 1946).

### 3.4. Medicinal Uses

Pigeon pea has several uses as medicine. It is used in Ayurveda as volerant, a medicine that heals wounds and sores, as an astringent, a medicine that stops bleeding by constricting the tissues, and as a medicine that cures diseases of the lungs and chest. It also works as antihelminthic to destroy internal worms (Faris and Singh 1990). Ekeke and Shode (1985) reported that pigeon pea causes reversion of sickled cells in patients suffering from sickle-cell anaemia. Pigeon pea leaf extract has some anti microbial activity towards the human pathogen (Okigbo and Omodamiro 2007). Till date only small fraction of study has been done for medicinal purpose use of pigeon pea, only extraction based study has been done that shows the further scope of dig into finding the medicinal use of this pulse corp. Pigeon pea is an integral part of traditional folk medicine in India,

China, and some other nations. It is known to prevent and cure human ailments like bronchitis, coughs, pneumonia, respiratory infections, pain, dysentery, menstrual disorders, curing sores, wounds, abdominal tumors, and diabetes in traditional folk medicine.

**Table 2. Worldwide ethnomedical uses of pigeon pea**

| Sr. no. | Country | Uses |
|---|---|---|
| 1. | Argentina | Bronchitis, coughs, genital irritation, pneumonia, skin problems |
| 2. | Brazil | Blood disorders, coughs, fevers, inflammation, pain, respiratory infections, sores, ulcers |
| 3. | China | Antidote, expectorant, sedative, vermifuge, vulnerary; for tumors |
| 4. | Cuba | Bronchitis, colds |
| 5. | Dominican Republic | Chest problems, sores, sore throat, wounds |
| 6. | Haiti | Antidote (Manihot), gargle, and vulnerary; for jaundice, urticaria, wounds |
| 7. | India | Colic, convulsions, leprosy, tumors (abdomen) |
| 8. | Malaysia | Abdominal pain, coughs, dermatosis, diarrhea, earache, enteritis, sores |
| 9. | Mexico | Astringent, diuretic, laxative, vulnerary; for dysentery |
| 10. | Peru | Anemia, diabetes, dysentery, hepatitis, menstrual disorders, urinary infections, yellow fever; as a diuretic |
| 11. | Trinidad | Flu, strokes |

Source: Singh, 2016.

## 4. Nutrition Value

Pigeon pea is a rich source of food proteins that is generally grown under risk-prone marginal lands. It occupies an important place among pulses and has been rated the best as far as its biological value is concerned.

**Table 3. Nutrient content of pigeon pea green seed, mature seed and processed dal**

| Constituents | Green seed | Mature seed | Dal |
|---|---|---|---|
| Protein (%) | 21.0 | 18.8 | 24 |
| Protein digestibility (%) | 66.8 | 58.5 | 60.5 |
| Trypsin inhibitor (units $mg^{-1}$) | 2.8 | 9.9 | 13.5 |
| Starch (%) | 48.4 | 53.0 | 57.6 |
| Starch digestibility (%) | 53.0 | 36.2 | - |
| Amylase inhibitor (units $mg^{-1}$) | 17.3 | 26.9 | - |
| Soluble sugars (%) | 5.1 | 3.1 | 5.2 |
| Flatulence factors (g 100 g- 1 soluble sugar) | 10.3 | 53.5 | - |
| Crude fiber (%) | 8.2 | 6.6 | 1.2 |
| Fat (%) | 2.3 | 1.9 | 1.6 |
| Minerals and trace elements (mg 100 g dry matter) | | | |
| Calcium | 94.6 | 120.8 | 16.3 |
| Magnesium | 113.7 | 122.0 | 78.9 |
| Copper | 1.4 | 1.3 | 1.3 |
| Iron | 4.6 | 3.9 | 2.9 |
| Zinc | 2.5 | 2.3 | 3.0 |
| Vitamins (mg $100^{-1}$ g fresh weight of edible portion) | | | |
| Carotene (vit A $100^{-1}$ g) | 469.0 | - | - |
| Thiamin (vit $B_1$) | 0.3 | - | - |
| Riboflavin (vit $B_2$) | 0.3 | - | - |
| Niacin | 3.0 | - | - |
| Ascorbic acid (vit C) | 25.0 | - | - |

Faris et al., 1987.

It has been recommended for a balanced diet with cereals, especially to fill in the nutritional gap for proteins among the poorer section in developing economies that cannot afford a nonvegetarian diet. At present, the protein availability in developing countries is about one third of normal requirements, and with ever-growing population, various nutritional development programs are facing a tough challenge to meet the protein demand. In general, pigeon pea can be grown both as annual crop or perennial plants in homestead and is consumed either as decorticated splits or green seeds as vegetables. It has been found that vegetable pigeon pea is considered superior to dry splits in crude fiber, fat, protein digestibility, as well as trace elements and minerals. Besides its nutritional value, pigeon pea also possesses various medicinal properties due to the presence of a number of polyphenols and flavonoids.

**Table 4. Availability of essential amino acids in pigeon pea seeds**

| Essential amino acid | Available mg/g of protein | Min. required mg/g of protein |
|---|---|---|
| Tryptophan | 9.76 | 7 |
| Threonine | 32.34 | 27 |
| Isoleucine | 36.17 | 25 |
| Leucine | 71.3 | 55 |
| Lysine | 70.09 | 51 |
| Methionine + cystine | 22.7 | 25 |
| Phenylalanine + tyrosine | 110.4 | 47 |
| Valine | 43.1 | 32 |
| Histidine | 35.66 | 18 |

Singh, 2016.

Pigeon pea seed is composed of cotyledons (85%), embryo (1%), and seed coat (14%). In contrast to the mature seeds, the immature seeds are generally lower in all nutritional values; however, they contain a significant amount of vitamin C (39 mg per 100 g) and have a slightly higher fat content. Research has shown that the protein content of the immature seeds is of a high quality (Singh, 2016).

## 4.1. Nutritional Value of Immature (Vegetable) Seeds

Green pigeon pea seeds (vegetable pigeon pea) are considered superior to dry splits in nutrition. The observations recorded at ICRISAT showed that pigeon pea dal was better than green peas with respect to starch and protein (Table 3). The green pigeon pea seeds contains higher crude fiber, fat, and protein digestibility. As far as trace and mineral elements are concerned, the green pea was found better in phosphorus by 28.2%, potassium by 17.2%, zinc by 48.3%, copper by 20.9%, and iron by 14.7%. The dal, however, had 19.2% more calcium and 10.8% more manganese (Faris and Singh 1990). Singh et al. 1977 reported that the vegetable-type pigeon pea had high polysaccharides and low crude fiber content than dal, irrespective of their seed sizes. They also reported that crude fiber contents in vegetable pigeon pea and garden pea (*Pisum sativum*) were comparable. There was a vast range in size and color in immature pods and mostly the consumer preference was for large green pods; however, these traits were not related to any organoleptic property of seeds.

## 4.2. Nutritional Changes in Developing Seeds

From nutritional and marketing points of view, it is essential that the green pods are picked at a proper stage to reap maximum seed yield with highest nutritional quality. In this context, Singh et al. (1991) observed that in growing seeds, the starch content was negatively associated with their protein and sugar contents. The amount of crude fiber content in the growing seeds increased slowly with maturation. The soluble sugars and proteins decreased in the developing seeds, but the starch content increased rapidly between 24 and 32 days after flowering, showed that the minerals and trace elements such as calcium, iron, zinc, magnesium, and copper remained more or less constant and did not change markedly during seed development in pigeon pea (Meiners et al., 1976).

## 4.3. Nutritional Value of Split Peas (Dal)

Carbohydrates (67%) and protein (22%) are main constituents of pigeon pea seeds (Faris and Singh 1990). The quality of protein is determined by its quantity and digestibility and amino acid contents. In pigeon pea the amino acids such as lysine and threonines are in good proportions, while methionine and cystine are deficient (Singh and Jambunathan 1982). Pigeon pea cotyledons are also rich in calcium and iron. Wild relatives of pigeon pea are used as highprotein donor parents and it was demonstrated that seed protein content in cultivated types can be enhanced through conventional breeding. Singh et al. (1990) reported large differences between the levels of protein in high-protein (28.7–31.1%) lines and control cultivars (23.1–24.8%). As expected, the starch component in high-protein lines was relatively less (54.3–55.6%) than that of controls (58.7–59.3%). Also the high-protein lines were marginally lower (2.5–2.6%) in fat content when compared with control cultivars (2.9–3.1%). The differences in the major protein fractions of the high- and normal-protein lines were also large. In comparison to controls (60.3–60.5%), the globulin fraction was higher (63.5–66.2%) in the high-protein lines and the reverse was true for glutelin.

The white-seeded pigeon pea cultivars contain relatively less amounts of polyphenols. Such cultivars are preferred in many countries where dehulling facilities are not available and whole seeds are consumed. In comparison to the white seeded cultivars, the red-seeded types contain three times greater quantity of polyphenols (Singh 1984). Similarly, the enzyme inhibition activity was also greater in the colored seeds of pigeon pea. Since in India almost the entire pigeon pea production (3.2 m tones) is dehulled and converted into dal for consumption, the tannins present in the colored seed coat pose no nutritional problem.

## 4.4. Nutrient Losses during Storage and Processing

Pigeon pea is predominantly cultivated by small holder farmers, and they generally store the whole seeds for over 6–12 months round the year for consumption. In rural areas the farmers process small quantities of grain as and when needed, and they do dehulling by hand-operated traditional grinding stones called chakki or quern. Since in pigeon pea the cotyledons are attached tightly with seed coat by various gums, the dehulling involves the process of dissolving the gum layers by soaking whole seeds in water, heat treatment, or adding oil after surface scarification. This is followed by drying, dehusking, and splitting of cotyledons. During these processes, the losses of certain proportions of cotyledons are inevitable, and it is estimated that by using advanced processing technology about 15–17% of grain mass is lost, while with chakki such losses may shoot up to 20–25%. In rural areas, the seeds are generally stored in gunny bags or bins made of mud and husk (Singh, 2016).

Pigeon pea seeds when stored for 8 months turned unfit for consumption as their total uric acid content crossed the safe prescribed limit (Vimala and Pushpamma 1983). Cooking time of pigeon pea in general increased with storage time (Vimala and Pushpamma 1985). The storage of pigeon pea seeds also resulted in the loss of vitamins. Such losses were less (10–26%) in the protected seed and high (32–49%) in the infested seeds (Reddy and Pushpamma 1981). Thiamine and niacin contents also registered decline during storage. Factors such as moisture, temperature, relative humidity, and seed hardness determined the extent of quality losses during storage (Saxena et al. 2008). In comparison to inner layers of cotyledons, the outer layers are rich in protein (Reddy et al., 1979). From a nutritional point of view, this is a matter of concern since dehulling not only removes protein-rich germ but also some proportion of the outer layers of the cotyledons. Singh and Jambunathan (1990) further observed that dehulling also removes about 20% calcium and 30% iron. According to Kurien (1981), the dal yield under controlled conditions achieves an efficiency of 80–84%, but at commercial level the recovery remains around 70%. Therefore, with a combination of a superior variety and efficient processing technology, the nutrient availability can be maximized.

## 4.5. Chemical Constituents of Leaves

In order to know the major chemical constituents of pigeon pea leaves, efforts were made to isolate and identify various active chemical compounds. The research efforts revealed that some polyphenols, especially flavonoids, play an important role in curing certain human ailments. The four major flavonoids identified in the extracts of pigeon pea leaves are quercetin, luteolin, apigenin, and isorhamnetin. These compounds are known for their important pharmacological activities (Chen et al. 1985; Green et al. 2003; Li et al. 1999). Flavonoids are polyphenolic compounds, which are widely found in plant kingdom. As intrinsic components of fruits, vegetables, and beverages, many of the 4000 different flavonoids known to date are present in a common regular diet (Crozier et al. 1997). Pigeon pea leaves also contain other compounds such as hordenine, juliflorine, betulinic acid, stigmasterol, beta-sitosterol, etc. In recent years, extensive research is being carried on various antibacterial, antifungal, antiviral, anticancer, and anti-inflammatory properties of these flavonoids (Matsuda et al. 2003; Srinivas et al. 2003; Kim et al. 1999).

## CONCLUSION

Pigeon pea is excellent pulse crop with different usage and high nutritional value. The continuous breeding efforts to improve crop production and nutrient avability makes it optimistic for future use especially in the developing countries like African continent for children and pregnant women to eradicate malnutrition.

## REFERENCES

Akhtaruzzaman, A. F. M., Siddique, A. B., & Chowdhury, A. R. (1986). Potentiality of pigeon pea (Arhar) [*Cajanus cajan* of Bangladesh] plant for pulping. *Bano Biggyan Patrika (Bangladesh)*, 15(1-2): 31-36.

Chen, D. H., Li, H. Y., & Lin, H. (1985). Studies on chemical constituents in pigeonpea leaves. *Chin. Tradit. Herb Drugs*, 16: 134-136.

Crozier, A., Jensen, E., Lean, M. E., and McDonald, M. S. (1997). Quantitative analysis of flavonoids by reversed-phase high-performance liquid chromatography. *Journal of Chromatography A*, 761 (1-2): 315-321.

Damardjati, D. S., & Widowati, S. (1985). Prospects on development of pigeonpea in Indonesia. *Journal Penelitian dan Pengembangan Pertanian*, 3: 53-59.

Ekeke, G. I., & Shode, F. O. (1985). The reversion of sickled cells by *Cajanus cajan*. *Planta medica*, 51(06): 504-507.

Faris, D. G., & Singh, U. (1990). Pigeonpea: nutrition and products. *Pigeonpea: nutrition and products.*, 401-433.

Faris, D. G. (1987). *Vegetable pigeonpea: a promising crop for India* (No. Folleto 11554).

Ganapathy, K. N., Gnanesh, B. N., Gowda, M. B., Venkatesha, S. C., Gomashe, S. S., & Channamallikarjuna, V. (2011). AFLP analysis in pigeonpea (*Cajanus cajan* (L.) Millsp.) revealed close relationship of cultivated genotypes with some of its wild relatives. *Genetic Resources and Crop Evolution*, 58 (6): 837-847.

Green, P. W., Stevenson, P. C., Simmonds, M. S., & Sharma, H. C. (2003). Phenolic compounds on the pod-surface of pigeonpea, *Cajanus cajan*, mediate feeding behavior of *Helicoverpa armigera* larvae. *Journal of Chemical Ecology*, 29 (4): 811-821.

ICRISAT (International Crops Research Institute for the Semi-Arid Tropics). (1984). *Annual report 1983*. Patancheru, A. P. 502 324, India: ICRISAT.

Jain, K. C., Faris, D. G., & Reddy, M. C. (1987). Performance of medium-duration pigeonpea genotypes for wood and grain yield in pigeonpea. *International Pigeonpea Newsletter*, 6: 34-35.

Kim, H. K., Cheon, B. S., Kim, Y. H., Kim, S. Y., & Kim, H. P. (1999). Effects of naturally occurring flavonoids on nitric oxide production in the macrophage cell line RAW 264.7 and their structure–activity relationships. *Biochemical pharmacology*, 58(5): 759-765.

Kulkarni, N. K., Reddy, A. S., Kumar, P. L., Vijaynarasimha, J., Rangaswamy, K. T., Muniyappa, V., Reddy, L. J., Saxena, K. B., Jones, A. T. and Reddy, D. V. R. (2003). Broad-based resistance to pigeonpea sterility mosaic disease in accessions of *Cajanus scarabaeoides* (L.) Benth. *Indian Journal of Plant Protection*, 31 (1): 6-11.

Kumar, P. L., Jones, A. T., Sreenivasulu, P., & Reddy, D. V. R. (2000). Breakthrough in the Identification of the Causal Virus of Pigeonpea Sterility Mosaic Disease. *Journal of Mycology and Plant Pathology*, 30 (2): 249-249.

Kumar, P. L., Jones, A. T., & Reddy, D. V. R. (2003). A novel mite-transmitted virus with a divided RNA genome closely associated with pigeonpea sterility mosaic disease. *Phytopathology*, 93 (1): 71-81.

Kurien, P. P. (1980, December). Advances in milling technology of pigeonpea. In *International Workshop on Pigeonpeas*, 1: 321-328.

Li, L., Ning, X., & Zihua, C. (1999). Flavonoids from *Cajanus cajan* L. *Journal of China Pharmaceutical University*, 30 (1): 21-23.

Mallikarjuna, N., & Saxena, K. B. (2005). A new cytoplasmic nuclear male-sterility system derived from cultivated pigeonpea cytoplasm. *Euphytica*, 142 (1-2): 143-148.

Matsuda, H., Morikawa, T., Ando, S., Toguchida, I., & Yoshikawa, M. (2003). Structural requirements of flavonoids for nitric oxide production inhibitory activity and mechanism of action. *Bioorganic & medicinal chemistry*, 11 (9): 1995-2000.

Meiners, C. R., Derise, N. L., Lau, H. C., Crews, M. G., Ritchey, S. J., & Murphy, E. W. (1976). The content of nine mineral elements in raw and cooked mature dry legumes. *Journal of Agricultural and Food Chemistry*, 24 (6): 1126-1130.

Morton, J. F. (1976). The pigeon pea (*Cajanus cajan* Millsp.): a high protein tropical bush legume. *HortScience*, 11 (1): 11-19.

Okigbo, R. N., & Omodamiro, O. D. (2007). Antimicrobial effect of leaf extracts of pigeon pea (*Cajanus cajan* (L.) Millsp.) on some human pathogens. *Journal of herbs, spices & medicinal plants*, 12 (1-2): 117-127.

Patel, P. T., & Tikka, S. B. S. (2014). Gene action and stability parameters for yield and yield components, maturity duration and protein content of CGMS lines, pollen fertility restorers and their hybrids in pigeonpea [*Cajanus cajan* (L.) Millsp.]. *Euphytica*, 199 (3): 349-362.

Pathak, G. N. (1970). Red gram. *Pulse crops of India*, 14-53.

Reddy, L. J., Green, J. M., Singh, U., Bisen, S. S., & Jambunathan, R. (1979). Seed protein studies on *Cajanus cajan*, *Atylosia* spp. and some hybrid derivatives [pigeon peas]. In *International Symposium on Seed Protein Improvement in Cereals and Grain Legumes.* Neuherberg (Germany, FR). 4 Sep 1978.

Reddy, M. V., & Nene, Y. L. (1980, December). Estimation of yield loss in pigeonpea due to sterility mosaic. In *Proceedings of International workshop on pigeonpeas*, 2: 15-19.

Reddy, M. V., Sharma, S. B., & Nene, Y. L. (1990). Pigeonpea: disease management. *Pigeonpea: disease management.*, 303-347.

Saxena, K. B. (2008). Genetic improvement of pigeon pea-a review. *Tropical plant biology*, 1 (2): 159-178.

Saxena, K. B., Kumar, R. V., Singh, L., & Raina, R. (1997, March). Development of a cytoplasmicnuclear male-sterility system in pigeonpea. In *Progress report fourth consultative group meeting on cytoplasmic male-sterility in pigeonpea*, 3-4.

Saxena, K. B., Kumar, R. V., Srivastava, N., & Shiying, B. (2005). A cytoplasmic-nuclear male-sterility system derived from a cross between *Cajanus cajanifolius* and *Cajanus cajan*. *Euphytica*, 145 (3): 289-294.

Saxena, K. B., Ravikoti, V. K., Dalvi, V. A., Pandey, L. B., & Gaddikeri, G. (2010). Development of cytoplasmic–nuclear male sterility, its inheritance, and potential use in hybrid pigeonpea breeding. *Journal of Heredity*, 101 (4): 497-503.

Sheahan, C. M. (2012). *Plant guide for pigeon pea (Cajanus cajan).* USDA-Natural Resources Conservation Service, Cape May Plant Materials Center. Cape May, NJ, 8210.

Singh, A. K., Rathi, Y. P. S., & Agrawal, K. C. (1999). *Sterility mosaic of pigeonpea: A challenge of 20$^{th}$ Century*, 15: 85-92.

Singh, I. P. (2016). Nutritional benefits of pigeonpea. In *Biofortification of Food Crops*, Springer, New Delhi, 73-81.

Singh, Laxman, N. Singh, M. P. Shrivastava, and A. K. Gupta. (1977). "Characteristics and utilization of vegetable types of pigeon peas

(*Cajanus cajan* (L.) Millsp.)." *Indian journal of nutrition and dietetics*, 14: 8-14.

Singh, U. (1984). The inhibition of digestive enzymes by polyphenols of chickpea (*Cicer Arietinum* L.) and pigeonpea (*Cajanus Cajan* (L.) Millsp.). *Nutrition reports international*, 29 (3): 745-753.

Singh, U., & Jambunathan, R. (1982). Distribution of seed protein fractions and amino acids in different anatomical parts of chickpea (*Cicer arietinum* L.) and pigeon pea (*Cajanus cajan* L.). *Plant Foods for Human Nutrition*, 31 (4): 347-354.

Singh, U. and Jambunathan, R., (1990). Pigeon pea: postharvest technology. *Pigeon pea: postharvest technology.*, 435-455.

Singh, U., Rao, P. V., Saxena, K. and Singh, L., (1991). Chemical changes at different stages of seed development in vegetable pigeonpeas (*Cajanus cajan*). *Journal of the Science of Food and Agriculture*, 57 (1): 49-54.

Singh, U., Jambunathan, R., Saxena, K. and Subrahmanyam, N., (1990). Nutritional quality evaluation of newly developed high-protein genotypes of pigeon pea (*Cajanus cajan*). *Journal of the Science of Food and Agriculture*, 50 (2): 201-209.

Squibb, R. L., Falla, A., Fuentes, J. A. and Love, H. T., (1950). Value of Desmodium, Pigeon pea fodder, and Guatemalan and United States alfalfa meals in rations for baby chicks. *Poultry Science*, 29 (4): 482-485.

Srinivas, K. V. N. S., Rao, Y. K., Mahender, I., Das, B., Krishna, K. R., Kishore, K. H. and Murty, U. S. N., (2003). Flavanoids from *Caesalpinia pulcherrima*. *Phytochemistry*, 63 (7): 789-793.

Tikka, S. B. S., Parmar, L. D. and Chauhan, R. M., (1997). First record of cytoplasmic Genit male sterility system in pigeonpea (*Cajanus cajan* (L.) Millsp.) through wide hybridization. *Gujarat Agricultural University Research Journal*, 22: 160-162.

U ma Reddy, M. U. and Pushpamma, P., (1980). December. Effect of insect infestation and storage on the nutritional quality of different varieties of pigeon pea. In *International Workshop on Pigeonpeas*, 443–451.

Upadhyaya, H. D., Sharma, S., Reddy, K. N., Saxena, R., Varshney, R. K. and Gowda, C. L., (2013). Pigeon pea. In *Genetic and genomic resources of grain legume improvement*, 181-202.

Van der Maesen, L. J. G., P. Remanandan, and Anishetty N. Murthi. (1980) "Pigeon pea genetic resources." In *International Workshop on Pigeon peas*, 385.

Vimala V, Pushpamma P. (1983). "Storage quality of pulses stored in three agroclimatic regions of Andhra Pradesh-quantitative changes." *Bull Grain Technol*, 21: 54–62.

Vimala, V. and Pushpamma, P., (1985). Effect of improved storage methods on cookability of pulses stored for one year in different containers. *Journal of food science and technology (Mysore)*, 22 (5): 327-329.

Wallis, E. S., Woolcock, R. F. and Byth, D. E., (1988). *Potential for pigeonpea in Thailand, Indonesia and Burma.*

Wanjari, K. B., Patil, A. N., Manapure, P., Manjayya, J. G. and Patel, M., (1999). Cytoplasmic male sterility in pigeonpea with cytoplasm from *Cajanus volubilis*. *Annals of Plant Physiology*, 13 (2): 170-174.

# INDEX

## A



## B

## C



## D

## E



## F



## G

## H



## I

## J



## L



## M



## N



## O

## P



## Q



## R

## W